HANS Hirschberger

NEW DIRECTIONS IN TREE FRUIT PEST MANAGEMENT

NEW DIRECTIONS IN TREE FRUIT PEST MANAGEMENT

KATHLEEN WILLIAMS
General Editor

GOOD FRUIT GROWER
Yakima, Washington

Cover photograph: Adult *Pandemis* leafroller
courtesy of Jay Brunner, Washington State University,
Wenatchee, Washington.

Published by:
Good Fruit Grower, a division of
Washington State Fruit Commission
1005 Tieton Drive, Yakima, WA 98902
509/575-2315 (phone)
509/453-4880 (fax)

Library of Contress Cataloging-in-Publication Data
New directions in tree fruit pest management /
Kathleen Williams, general editor
p. cm.
"Proceedings of a short course...held in February
1991"–Introduc.
Includes bibliographical references.
ISBN 0-9630659-0-4 : $10.00
1. Fruit–Diseases and pests–Control–Northwest,
Pacific. 2. Fruit–Diseases and pests–Integrated
control–Northwest, Pacific. 3. Fruit–Diseases and
pests–Northwest, Pacific. I. Williams, Kathleen M.,
1951- . II. Good Fruit Grower (Wash.). III. Title:
Tree fruit pest management.
SB608.F8N38 1991
634'.0496'09795—dc20 91-73737
CIP

Printed and bound in the United States of America
10 9 8 7 6 5 4 3 2 1

CONTENTS

CONTENTS

Part 2

FUNDAMENTAL PRINCIPLES OF INTEGRATED PEST MANAGEMENT

Part 3

BIOLOGY & CONTROL OF SELECTED TREE FRUIT PESTS

INTRODUCTION

THE U.S. TREE FRUIT INDUSTRY IS UNDER INCREASING pressure to grow fruit with fewer chemical inputs. Since the late 1980's, the number of chemicals available to fruit growers has decreased substantially. Some of the important tools that growers have lost because the manufacturers did not pursue re-registration include daminozide, elgetol, and phosphamidon. The list of chemicals that are still in the decision-making process is a long one, and includes many uses that may be critical to some operations. Registrants can only re-register current uses (or register new uses) if that use will be profitable, and tree fruits are minor crops. We can look forward to higher costs for our crop protection chemicals in the future, as well as fewer chemicals with which to work.

The recent push on the part of the U.S. Environmental Protection Agency to re-register older compounds reflects a renewed public interest in a number of issues associated with pesticide use. These include residues on foods and their possible effect on human health; worker safety and re-entry standards; pesticide disposal; and environmental impact. In addition to the concerns that affect society as a whole, there are some additional issues that are of immediate concern to agriculture. These include resistance to pesticides by arthropod and disease pests (e.g., mites and Plictran; fireblight, and streptomycin), and destruction of beneficial organisms by broad-spectrum pesticides. Both sides of these issues point to the same direction: fuller implementation of integrated pest management programs, and the use of new pest control technologies less hazardous to nontarget organisms (including humans).

The concept of IPM has been around in a more or less formalized manner for three decades, but not all of the original hopes have been fulfilled. Some of the success stories have been striking, such as the integrated mite control currently prevailing in Washington State apple orchards. In other cases, we have not gone much beyond more precise timing of chemical control measures (as in the case of codling moth). The essential concepts of sampling a pest or disease population, and making a decision based on an economic threshold have not been implemented as fully as they might be. In the meantime, the array of pests we have to deal with has expanded in the past 10 years. Leafhoppers, leafminers, and leafrollers have become more serious in more areas than ever before, and mealybugs and Campylomma problems are more widespread than ever. Overhead sprinklers have aggravated scab problems, and the outbreak of fireblight in 1988 devastated many pear-growing areas. As we attempt to expand our markets overseas, or even within the United States, the question of phytosanitary standards makes pest control issues more critical.

This book is the proceedings of a shortcourse on tree fruit pest management held in February 1991, and the choice of topic was motivated by the concerns discussed above. The shortcourse and the proceedings are part of an ongoing effort to make tree fruit growers aware of the latest information and technology. Many thanks are due to the authors who graciously shared their time and expertise with the Pacific Northwest tree fruit industry.

KATHLEEN M. WILLIAMS
General Editor
ELIZABETH H. BEERS
Entomology Editor
GARY G. GROVE
Plant Pathology Editor

NEW DIRECTIONS IN TREE FRUIT PEST MANAGEMENT

Part 1

Tree Fruit Diseases: Life Cycles & Control Measures

1

DISEASE MANAGEMENT STRATEGIES FOR CONTROLLING APPLE SCAB

MICHAEL A. ELLIS
The Ohio State University
Ohio Agricultural Research
and Development Center
Wooster, Ohio

SCAB IS CAUSED BY THE FUNGUS *VENTURIA INAEQUALIS*, which occurs in most areas of the world where apples are grown. It is less severe in semi-arid regions than in cool, humid areas with frequent rainfall. The climate in all the fruit-growing regions east of the Rockies is extremely favorable for scab. The drier climates in many of the western apple production regions are generally much less conducive to scab development.

Primary Disease Cycle

The fungus overwinters in leaves on the orchard floor. In late fall and early spring, microscopic, black, pimple-like structures, called perithecia, are produced in these dead leaves. Within each perithecium are asci, each with eight ascospores. The ascospores produce the first, or primary, infections on the new growth in the spring.

Perithecial development in the old dead leaves is favored by alternating periods of wetness and dryness in late winter and early spring. Normally, some perithecia have mature ascospores when the blossom buds start to open. When the leaves on the orchard floor become wet, spores are forcibly ejected into the air. Air currents carry them to the emerging tissues where infection takes place. Maturation and discharge of ascospores usually lasts five to nine weeks. After this time, there are no ascospores left to cause primary infections.

Ascospore germination begins as soon as a spore lands on new, green leaves or fruit, provided a film of moisture is present. If primary infections occur, lesions are produced on leaves and fruit. Each lesion (which was caused by one ascospore) will produce hundreds of thousands of conidia, each of which is capable of causing a new infection. The number of hours of wetting required for infection varies with prevailing temperatures. Information on the environmental conditions (leaf wetness duration and temperature) required for scab infection is provided in Washington State University Cooperative Extension Bulletin EB1044.

It is important that growers understand the basic disease cycle for apple scab in order to develop the most successful disease management program.

The Bottom Line for Disease Control

In order to effectively control apple scab, you *must* control primary infections from ascospores. If ascospores are prevented from infecting leaves and fruit early in the season, no further scab control measures are needed after the supply of ascospores is depleted. However, if early-season infections are not controlled, additional fungicide protection is needed throughout summer to protect against secondary infections by conidia. The number of conidia produced by just a few early-season scab lesions is much greater than the total number of ascospores produced in an entire acre of leaf litter in most clean, commercial orchards.

APPLE SCAB FUNGICIDES AND TIMING SPRAYS FOR PRIMARY SCAB CONTROL

Fungicides

The fungicides available for scab control can be broadly classified as follows:

1) Protectants: Captan; Thiram/ Ziram; Ferbam; Sulfur; Copper. These fungicides protect fruit and foliage by preventing spore germination. They will not arrest lesion development after infection occurs.

2) Benzimidazoles and dodine: Benlate, Topsin-M, Syllit (dodine). These fungicides have a combination of protectant and eradicant activity and have limited systemic activity. They will provide some "burn out" of scab infections after they occur. They have very limited curative or "kickback activity."

3) Sterol-inhibiting (SI) fungicides: Rubigan and Rally. These fungicides provide good curative or "kickback activity." They provide at least 72 hours of kickback, and as much as 96 hours. This means they can be applied 72 to 96 hours after the initiation of an infection period and still provide control of apple scab. The sterol-inhibiting fungicides are generally not considered to have strong protectant activity.

Timing

1) Standard protectant program (7-day interval). Protectant fungicide must be applied at intervals of approximately seven days during primary scab, because growing clusters and terminals are constantly producing new tissues that need protection. The old rule-of-thumb is "Repeat protectant sprays after seven days of new growth or one inch of rain."

2) Extended protectant program (10-day interval—not to exceed 14 days). If a combination of protectant and sterol-inhibiting fungicides is used, the spray interval can be extended to 10 days. In practice, the seven-day protectant and 10-day extended protectant schedules can be modified to take advantage of weather predictions. If a protectant program is used and no rain is predicted on days seven, eight, and nine after the last application, the next application can be made just before the next rain. The sterol-inhibiting/protectant combination provides approximately five to six days of protection and four to five days of curative activity.

Using the extended protectant program, the interval can be extended one day for each day without infection, on days six through nine after the last application. With sterol-inhibiting/protectant combinations, the interval should never exceed 14 days, because some lesions partially inactivated by the sterol-inhibiting fungicides may not be completely controlled without a second spray. The extended protectant program is being adapted by many apple growers in the Midwest and Northeast. This program should result in fewer sprays than the standard protectant program.

3) Postinfection or curative program. The sterol-inhibiting fungicides can also be used on a postinfection schedule, where fungicides are applied only after infection periods occur. This approach should also reduce the number of fungicide applications

needed per season, especially in dry growing seasons. By waiting to spray until after a scab infection period actually occurs, growers can avoid extra sprays that would be applied if they sprayed according to weather predictions.

Growers who do not have the capability to accurately monitor infection periods should not use a postinfection program. Another problem with this approach is extended rainy periods that can interfere with postinfection applications that must be made within 72 to 96 hours after the initiation of an infection period.

CONTROLLING SECONDARY SCAB

Secondary Disease Cycle

Secondary infections are initiated by conidia produced in primary lesions. Since conidia may develop as soon as seven to nine days after infection, secondary infections may be initiated very early in the growing season (prebloom). This is particularly true when ascospores infect the apical portions of sepals and leaves at the silver tip stage of bud development. Conidia are disseminated by splashing rain and by wind. Conidial germination and infection occurs under about the same conditions as for ascospores.

Secondary infection of fruit can occur in the fall but not show up until the fruit have been stored for several months. The disease can also build up on the leaves after harvest. Since the fungus overwinters in these leaves, perithecia may be present in quantity to start the new season, even though a good spray program was followed the previous year.

Fungicides for Controlling Secondary Scab

If control of primary scab is not achieved and scab lesions develop, additional fungicide sprays are generally required on a 14-day interval throughout the remainder of the growing season, or until scab lesions are inactivated.

None of the currently registered fungicides are extremely effective in "burning out" established scab lesions. The benzimidazoles are probably the most effective for this purpose, but problems associated with the development of fungicide resistance prevent their prolonged use. The sterol-inhibiting fungicides have good curative (pre-symptom) activity, but are not highly effective in killing secondary scab spores in established lesions. In addition, the use of sterol-inhibiting fungicides for late season disease control (after primary scab) will probably be discouraged in the East because of: 1) their lack of control for summer diseases; 2) expense; and 3) the potential for fungicide resistance development due to excessive (season-long) use.

Captan is probably the next best fungicide for "burning out" established scab lesions. Especially during hot weather, Captan will inactivate scab lesions and is generally recommended for eradicant applications during mid to late summer.

The bottom line for control is to use a good early season program to control primary infections. Once infections are established in the orchard, the use of additional fungicide sprays throughout the growing season to control secondary scab will probably be required. There is no fungicide currently available that will totally eradicate secondary scab lesions, especially in wet, cool growing seasons.

APPROACHES TOWARD REDUCING THE USE OF FUNGICIDES FOR SCAB CONTROL

1) Use scab-resistant cultivars.

2) Use an extended protectant or postinfection spray program. Especially in dry growing seasons, these programs should result in less fungicide applications than a standard protectant program.

3) Delay the initiation of the fungicide program. In clean orchards with no or very little history of scab the previous season, delay the first scab spray until tight cluster. In several areas of the Northeast, growers in low-inoculum orchards are using this program. They delay the first spray until tight cluster, then make at least three applications with a sterol-inhibiting fungicide between tight cluster and petal fall. In orchards that had severe scab the previous season, the fungicide program needs to be initiated at green-tip or half-inch green at the latest.

4) Reduce scab inoculum levels in the orchard: a) after leaves drop in the fall, pulverize them with a mower. This will speed up their decay. Obviously, if leaves do not overwinter in the orchard, they cannot serve as a source of primary inoculum in the spring; b) apply nitrogen to leaves on the ground to speed up their degradation. In theory this sounds good, but I have not seen any recommended nitrogen rates for this purpose. I would not recommend this approach until more research has been conducted and specific recommendations are available.

5) Late season applications of benzimidazole (Benlate) fungicides may be beneficial in preventing ascospore formation in scab-infected leaves. This is another area that needs to be researched further.

6) Good pruning practices to open up the tree canopy will be beneficial in increasing the efficacy of the fungicide spray program.

Open canopies result in better air circulation and light penetration that will aid in reducing the length of wetness periods. They also allow for better spray penetration and coverage.

7) Properly calibrate the sprayer and make sure that you are getting complete coverage of all susceptible plant parts. This will become increasingly important as we use the newer fungicides with very low rates (5 oz. per acre).

8) Adjust irrigation practices. Especially when overtree irrigation is used, growers need to carefully examine their irrigation practices in relation to the apple scab disease cycle and disease development. Growers in the arid areas of Washington State who use overhead irrigation systems should consider the following points regarding the adjustment of irrigation practices to avoid creating infection periods.

Example: If a grower uses a 12-hour set to put out three inches of water, a 12-hour wetness period is created in the orchard. This is generally sufficient wetness duration to result in an apple scab infection period. If the grower used two 6-hour sets spaced one day apart (24 hours dry time), three inches of water would still be applied, but no infection period has been created.

Growers can also adjust irrigation practices to induce ascospore (primary inoculum) discharge, without inducing apple scab infection periods. If growers are applying three inches of irrigation a week (a 12-hour set equals 3 inches of water), irrigation timing could be adjusted to create sufficient wetting to induce ascospore discharge and depletion, while still not creating infection periods.

Ascospore discharge is induced after dead leaves on the ground have been wetted for at least half an hour. After about three hours of wetness, the maximum rate of discharge will be reached. After about six hours of wetness, 75 percent of the mature ascospores are discharged. Recent research indicates that ascospore discharge may occur only during daylight.

Example: If a grower was applying three inches of water per week in a 12-hour set, this could be adjusted to two 6-hour sets spaced one day apart (24 hours dry time). This would permit maximum discharge of mature ascospores without inducing an infection period. If this practice was conducted at weekly intervals from half-inch green through bloom, growers could be reasonably sure that all primary inoculum (ascospores) was discharged by the end of bloom. If this can be accomplished without getting primary infections, then scab is effectively controlled.

Note:
Whereas this discussion of adjusting irrigation practices to aid in disease control seems logical, I am not aware of any published data that demonstrates the successful use of this control strategy.

Growers who wish to use this strategy for apple scab control should proceed with caution. A thorough understanding of the disease cycle is essential to using a disease management strategy such as this.

2

PHYTOPHTHORA ROOT AND CROWN ROTS OF DECIDUOUS FRUIT TREES

WAYNE F. WILCOX
Cornell University
New York State Agricultural Experiment Station
Geneva, New York

OVER THE LAST QUARTER CENTURY, PHYTOPHTHORA root rot and crown rot (collar rot) have become increasingly serious diseases of deciduous fruit trees throughout the world. The reasons for this are probably several, but include shifting rootstock preferences (e.g., from relatively resistant seedling apple rootstocks to the more susceptible clonals), expansion of some fruit operations onto land only marginally suited for such crops, and the increased availability and utilization of irrigation water in more arid regions. Additionally, improved diagnostic techniques have repeatedly shown that many trees once thought to be suffering from winter injury, "sour sap," or "wet feet," are actually dying as the result of Phytophthora root and crown rots.

SYMPTOMS

Diseased trees are most likely to be found where soil type, topography, climate, or irrigation scheme have contributed to significant periods of excessive soil moisture in the recent past. Aboveground symptoms are variable, but may include any or all of the standard signals indicating root or trunk injury: lack of terminal growth, sparse off-color foliage, premature autumn coloration, wilting, and collapse. Since these symptoms may have a number of causes, field diagnosis requires an examination of the potential "root" of the problem, with the aid of a shovel and knife.

After removing several inches of soil from around the base of declining (not dead) trees, cut through the outer bark of the exposed crown region. The inner bark should normally be white; however, on most trees with Phytophthora crown rot, this tissue will be orange-brown or red-brown while it is still relatively intact, eventually turning darker brown as it decays. There should be no signs of white fungus growth upon or beneath the bark before it begins to decompose. If such growth is consistently found, it is likely that Armillaria root rot or a similar (non-Phytophthora) disease is involved.

Crown rot is the most "dramatic" phase of disease caused by *Phytophthora* fungi on fruit trees; however, depending upon the rootstock and Phytophthora species involved, infection is sometimes confined to the woody roots and/or feeder roots away from the crown. (Note that when only root rot is involved, trees typically appear stunted and sick but are much less likely to collapse and die as they often do when crown rot occurs.) If crown rot is not detected on a symptomatic tree, next look for similar red-brown or orange-brown lesions in the inner bark of the primary woody roots extending away from the crown, and on the secondary roots that branch from the primaries. Finally, dig up some of the fine feeder roots out near the drip line of the tree and check to see if they are dead (black in the middle) or alive (white in the middle). Although the red- or orange-brown lesions found on the crown can provide a pretty confident field diagnosis of infection by *Phytophthora* fungi, and to a lesser extent, so can similar lesions on primary and secondary roots, blackened feeder roots only tell you that *something* is wrong, but not what. In any case, positive confirmation of field diagnoses can and should be made by a diagnostic laboratory as needed.

CAUSE

Phytophthora root and crown rots are caused by about a dozen related soil-borne fungi which belong to the genus *Phytophthora*. Although these different fungi may vary somewhat in their temperature requirements, tendency to attack roots versus crown tissue, or aggressiveness against different rootstocks, all have one important trait in common—they require extremely wet or saturated soil in order to cause significant disease losses.

The *Phytophthora* fungi persist in the soil primarily as dormant resting spores (oospores) or in an active vegetative state (mycelium) within infected tissue. When the soil is moist or wet, spores of a different type (sporangia) are formed as a result of germinating oospores in the soil or as direct outgrowths of the active mycelium in the infected root and crown tissues.

By far, the most important stage in the disease cycle occurs after these sporangia have formed and the soil becomes completely saturated with water, i.e., flooded or puddled. When this occurs, the contents of each sporangium become transformed into approximately 40 individual spores of yet another type (zoospores), which are the main infective units of the fungus. The zoospores are then expelled from the sporangia into the soil, where they use tail-like structures (flagella) to swim through the water-filled pore spaces in search of plant roots, to which they are chemically attracted. Zoospores may also swim to the surface water and be moved to other trees as it flows downhill. Contaminated surface water may also flow into streams, canals, or ponds, providing a mechanism for long-distance dispersal of the zoospores.

The influence of soil water conditions on these events cannot be overemphasized. For instance, it has been shown that whereas the maximum number of zoospores are released during periods of complete saturation or standing water (a reading of zero on an accurate tensiometer), only about 35 percent as many are released when that same tensiometer indicates a tension of 0.5 centibars (the equilibrium value in soil just 2 inches above a water table), only 10 percent as many zoospores are released at a tension of 1 centibar (the equilibrium value 4 inches above a water table); and virtually no zoospores are released at a tension of 2.5 centibars (10 inches above a water table). These same small changes in soil water tension have a similarly profound influence on the ability of zoospores to swim through the soil, since the very large pores through which they move are also those which drain at very low tensions.

Whether or not infection occurs once the zoospores are released into the soil and manage to reach the tree's roots or crown depends largely upon the genetic makeup of the rootstock and the physiological condition it is in. Although (too) many of the factors that influence a given tree's susceptibility to infection are not known, it appears that at least some rootstocks are less capable of resisting the pathogen's initial attempts at infection so long as flooded conditions continue and deprive them of oxygen. In effect then, periods of soil saturation which promote the production and dispersal of infective zoospores and furthermore reduce the tree's resistance to attack, serve as infection periods for these diseases. The minimum length of the saturated period necessary to produce an infection can be highly variable depending upon a number of different factors discussed below, but severity of the infection period is generally proportional to its duration. Some data from experiments conducted under greenhouse conditions are presented in Tables 1, 2, and 3 which illustrate these preceding concepts.

Some of the factors that interact with soil water conditions to determine the occurrence and severity of Phytophthora root and crown rots include: 1) the species of Phytophthora present and its (their) population level(s); 2) rootstock susceptibility; 3) soil temperature (cool to moderate temperatures are generally more favorable than high temperatures, but precise requirements vary among Phytophthora species); 4) tree age (there is some indication that trees become increasingly susceptible when they first come of bearing age); and 5) time of year (variable among rootstock x Phytophthora species combinations, but often trees are most susceptible within a few weeks on either side of bloom). Additional factors such as drought stress, excessive salinity, and nutritional status have been implicated as influencing the development of diseases caused by Phytophthora fungi on other crops, but these factors have received little, if any, study on deciduous fruit crops at present.

ORIGINS OF INOCULUM

Although some Phytophthora species appear possibly to be indigenous to particular fruit-growing locales (e.g., *P. cactorum* is commonly found in native woodlands in New York), others are almost certainly introduced onto individual farms. Sources of introducing inoculum include contaminated irrigation water (Phytophthora species that cause root and crown rots of deciduous fruit trees have been recovered from a number of irrigation canals and streams in western states and provinces); contaminated soil adhering to vehicles or machinery moving among orchards; and contaminated planting material (at least two major studies have shown this to be common). Whereas, it is desirable to exclude inoculum from individual orchards to whatever extent is possible, particularly in new orchard sites that may not already be contaminated, it is generally not practical to rely upon exclusion of these pathogens for control of the diseases that they cause.

CONTROL

Control of Phytophthora root and crown rots is best accomplished through an integrated program that involves: 1) a combination of management practices designed to deter water from standing or puddling around the tree crowns; 2) selection of tree species and rootstocks suitable for the drainage characteristics of the orchard; and, perhaps, 3) chemical control, if appropriate.

The importance of soil water management in controlling these diseases cannot be overemphasized. Making necessary site modifications to promote good drainage (tilling, ditching, ripping underlying pans or plow layers, etc.) has become second nature to most fruit growers, and for good reason. Berming, or planting on ridges, is a practice that can tilt the scales in the direction of increased tree survival in many marginally drained locations or in an unusually wet year, and should probably be practiced more than it currently is. Irrigation regimes that prevent prolonged periods of soil saturation are most likely to minimize losses from root and crown rots. For

TABLE 1

Effect of flooding duration and rootstock on severity of Phytophthora root and crown rots of cherry seedlings grown in a greenhouse.

		% Root rot		% Crown rot	
Pathogen	Hrs. flooded[1]	Mahaleb	Mazzard	Mahaleb	Mazzard
P. cryptogea	0	7	6	0	0
	12	23	20	0	0
	24	68	25	0	0
	48	99	92	5	4
P. megasperma	0	5	2	0	0
	12	10	6	0	0
	24	23	10	0	0
	48	87	81	1	0

[1] *Seedlings flooded for indicated periods once very two weeks*

NOTE: (1) Differences between Mahaleb and Mazzard were most pronounced when disease pressure was intermediate, e.g., 24-hr floodings with P. cryptogea. *Such differences were much less apparent when disease pressure was very high (48-hr floodings) or low (no flooding). (2)* P. cryptogea *was more destructive than* P. megasperma, *causing a more severe root rot, with intermediate flooding periods and a higher incidence of crown rot. (3) Both species caused crown rot only after the longest flooding treatment.*

example, frequent low-volume sprinkler applications are far less likely to promote disease development than is delivering the same amount of water in a few prolonged flood irrigations. If drip irrigating, carefully regulate output to avoid constant overwatering.

It can be dangerous to rank tree species and rootstocks for their general resistance to Phytophthora root and crown rots, because so many different factors combine to influence susceptibility to these

TABLE 2

Influence of soil oxygen levels on the development of Phytophthora root and crown rots of Mahaleb cherry seedlings when various treatments were imposed after inoculation with zoospores of *P. cryptogea*.

Post-inoculation treatment	Disease rating (0-5 scale)
Flooded 48 hr	4.2
Flooded 6 hr, then 42 hr @ 0.5% O_2	3.6
Flooded 6 hr, then 42 hr @ 2.5% O_2	1.7
Flooded 6 hr, then 42 hr @ 21% O_2 (air)	1.5

NOTE: *When seedlings were flooded for 6 hours following inoculation (to allow time for zoospores to swim and contact the roots and crown), disease was almost as severe if they were incubated for the next 42 hours at 0.5% soil oxygen—about 1/42 the level found in air—as when seedlings were flooded for the entire 48-hour period following inoculation. These results suggest that reduced oxygen levels in flooded soil contribute to the severity of Phytophthora root and crown rots that occur following flooded conditions.*

diseases. For instance, there appears to be a wide difference in susceptibility to *P. cactorum* among different apple rootstocks, whereas most rootstocks tested appear about equally susceptible to *P. cryptogea*. Similarly, the peak period of seasonal susceptibility is somewhat different for many individual Phytophthora species x rootstock combinations. Consequently, differential reactions among rootstocks may depend not only on which Phytophthora species is (are) present, but also on when periods of soil saturation occur during the year. With these qualifications in mind, however, here is how some of the major deciduous fruit tree species and rootstocks appear to stack up regarding their susceptibilities to these diseases, both in controlled tests and general field observations.

PEAR. Generally quite resistant. An important reason why you can grow pear trees in a site too wet for just about any other fruit crop.

APPLE. More susceptible than pears, but less susceptible than most stone fruits. Seedling rootstock is the most resistant, M.9, M.4, and M.2 are probably the best of the common clonals. Ottawa 3 also appears reasonably resistant. At the other end of the scale, MM.104 and MM.106 are notoriously susceptible, whereas M.7 and M.26 are probably somewhere in the middle. Opinion on MM.111 varies greatly, from relatively resistant to very susceptible; under New York conditions, it appears to be pretty comparable to M.7 and M.26. The interaction between rootstock clone, *Phytophthora* species, and flooding duration is illustrated by data from a greenhouse experiment summarized in Table 3.

PLUM. Most resistant of the stone fruits. Potential differences among various plum stocks is not yet clear, but none appear to be highly susceptible. John Mircetich (University of California, Davis) reports that Myrobalan appears to be more resistant than Mariana 2624.

CHERRY. Mahaleb, very susceptible; Mazzard and Morello, somewhat better; Colt, uncertain, but probably in the Mazzard/

TABLE 3

Influence of rootstock clone, *Phytophthora* species, and weekly flooding treatment on the number of apple liners (out of five tested) that developed crown rot under greenhouse conditions.

	P. cactorum Flooding duration				*P. cryptogea* Flooding duration			
Rootstock	0 hr	24 hr	48 hr	72 hr	0 hr	24 hr	48 hr	72 hr
M.7	2	1	1	4	0	0	2	4
M. 26	0	1	0	1	0	0	1	4
MM.111	0	1	2	5	0	0	3	4
Ottawa 3	0	0	1	2	0	0	2	5

	P. cambivora Flooding duration				*P. megasperma* Flooding duration			
	0 hr	24 hr	48 hr	72 hr	0 hr	24 hr	48 hr	72 hr
M.7	0	0	0	1	0	0	0	1
M.26	0	2	1	1	0	0	0	2
MM.111	0	0	1	3	0	0	1	3
Ottawa 3	0	0	1	2	0	0	0	0

NOTE: This is data from one greenhouse trial and should be interpreted as such. Nevertheless, it clearly shows the interactions between pathogen species, rootstock clone, and flooding duration on ultimate disease incidence.

Morello range. Some of the new candidate rootstocks being evaluated out of European and American programs look significantly more resistant than what is currently available, and may offer hope for the future. However, preliminary indications are that the new GM.9 and GM.79 rootstocks from Belgium should not be planted where Mahaleb will not survive, whereas GM.61 may have some useful resistance. Remember, though, that "preliminary indications" should be taken with a grain of salt—MM.106 was originally touted as a relatively resistant apple rootstock when it was first developed.

PEACH. Very susceptible. Range of suitable soils may be expanded if trees can be propagated on plum-type rootstocks. Mircetich reports that Nemaguard is more resistant than Lovell.

APRICOT. Very susceptible. As with peach, range of suitable soils may be expanded if trees can be propagated on plum-type rootstocks.

The discussion on chemical control has been saved for last, because this option should not be viewed as a control measure that reduces the need for good water management and appropriate rootstock selection. Again, experiences vary, but Ridomil and Aliette have been most effective when used as preventive treatments. These materials have consistently failed to revitalize trees already declining from crown rot (once a tree is girdled, it's girdled), although they may allow trees to come back if only root rot is involved. The decision of whether to implement chemical control procedures should be made on a case-by-case basis, influenced in part by the individual grower's philosophy on buying insurance.

3

POWDERY MILDEW OF CHERRY

GARY G. GROVE
Washington State University
Tree Fruit Research & Extension Center
Wenatchee, Washington

THE POWDERY MILDEWS ARE A UNIQUE GROUP of fungi that thrive in drier climates. Mildews affect a wide variety of crop plants. In our area, mildews are a perennial problem on apple, cherry, peach, apricot, and nectarine. Powdery mildew of cherry is caused by the fungus *Podosphaera clandestina*; powdery mildew of apples is caused by the closely related fungus *Podosphaera leucotricha*. Water is essential to many powdery mildews; however, they are not as a group as dependent upon water as, for example, the apple scab fungus. The powdery mildews are obligate parasites, i.e., parasites that can grow and multiply only on living organisms.

Powdery mildew of cherry fruit and foliage can be a severe problem in Washington State. The fruit phase of the disease is triggered by rains late in fruit development. Until recently, we had only one weapon to control the disease in Washington, and that was sulfur. Sulfur frequently fails to provide adequate control, due to its relatively short retention time, high disease pressure, or inferior spray coverage. We now have an additional weapon (Rubigan) in our arsenal, but it is expensive. In order to get the best disease control for our grower dollars, the life cycle of the mildew fungus had to be ascertained in order to optimize spray timings.

The symptoms of leaf infection are leaf curling or slight blistering. At times, a white moldy growth can be seen on the leaf surfaces. This consists of millions of spores of the fungus known as conidia. The fungus eventually forms small, circular fruiting structures (known as cleistothecia) on the leaf surface. Inside each cleistothecium are eight spores known as ascospores. The white moldy growth can also be observed on the surface of infected fruit. However, infected fruit may also show only a superficial water-soaked area.

Until recently, there was very little published information on the biology and life cycle of the cherry mildew fungus. Papers on the disease were purely descriptive in nature and consisted of reports from different geographical areas. Our research program on cherries has emphasized the biology of cherry mildew and the influence of weather conditions on outbreaks of the disease.

In order to unravel the pathogen's life cycle, we had to undertake a very thorough study on the fungus. First, we needed to determine the mode of overwintering. Powdery mildews overwinter as mycelium in dormant infected buds or as cleistothecia. Our experiments with the former have failed to provide any evidence that the fungus overwinters in buds. After overwintering in buds was ruled out, we undertook a large study on the cleistothecial stage. In order to demonstrate that the cleistothecium was the overwintering propagule, we needed to demonstrate that the ascospores survived the winter; that the cleistothecia would release ascospores when host tissue was present; and that the ascospores would germinate, infect, and be present in the orchard air prior to the appearance of mildew symptoms.

By periodically collecting cleistothecia from dead leaves on the orchard floor or trapped in tree crotches and visually assessing their spore contents, it was found that viability ranged from about 80 percent in mid January to 5 to 10 percent in mid May (*see Figure 1*). At bud burst, about 60 percent of the cleistothecia contained viable spores. When the cleistothecia were sprayed with water, they readily released ascospores—periodic collection and moistening of cleistothecia demonstrated that spore release occurred from mid March to mid May, with the largest spore release occurring in mid April (*see Figure 2*). Temperature and wetness period also had a significant effect on ascospore release—release occurred between 42

FIGURE 1

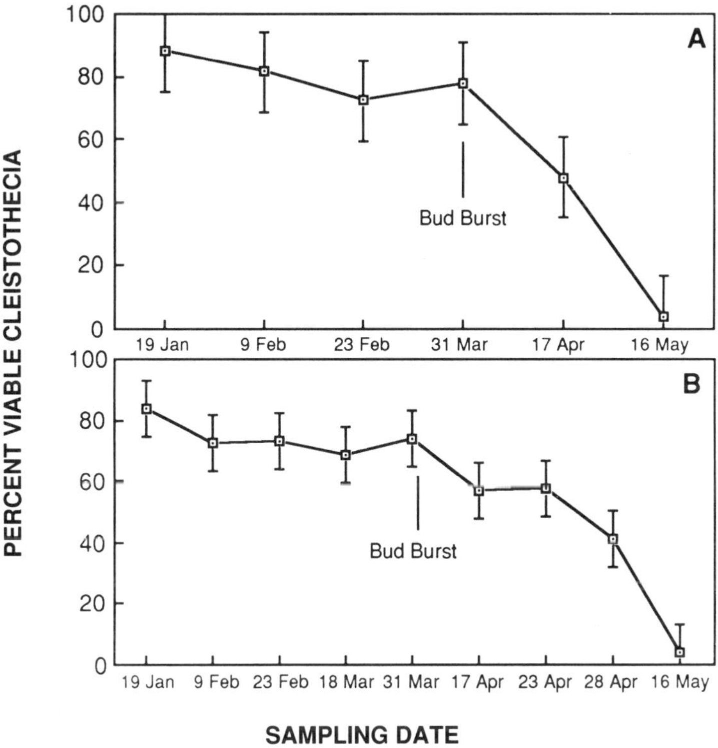

and 85 degrees Fahrenheit (°F) with an optimum temperature of 62°F. Release occurred at wetness periods as short as one hour, but in general required four to 24 hours to release in appreciable numbers. Temperature and wetness period also had a significant effect on ascospore germination—ascospore germination required eight-hour wetness periods at 78°F, and 16 to 24 hours at 62 and 70°F. By suspending cleistothecia over cherry seedlings, ascospore infectivity was demonstrated—infection occurred from early March to mid May *(see Figure 3)*. Inoculations made in mid April resulted in the highest mildew incidence and severity.

Ascospores were trapped from orchard air from early April to mid May, and were trapped prior to the appearance of mildew colonies and prior to the trapping of conidia (the secondary spore). Ascospores were trapped during rain or during wetness periods initiated by rain. The chronological sequence of ascospore trapping, appearance of primary mildew, and trapping of conidia was further evidence of cleistothecia being the principal primary inoculum source.

By considering the large amount of evidence that was gathered in the above studies, we concluded that cleistothecia were the principal mode of overwintering and that ascospores were the principal form of primary inoculum. More importantly, the presence of the primary inoculum in the orchard air occurred prior to the initial fungicide application. But perhaps the most important fact is the presence of a functional sexual stage in the disease cycle—a

FIGURE 2

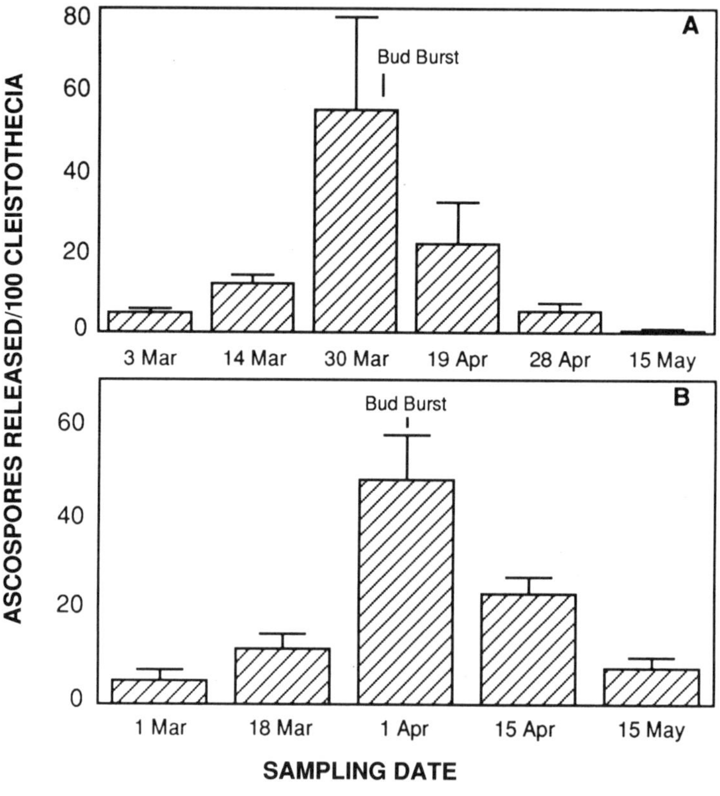

functional sexual stage provides the fungus with an easy means of varying genetically and a means to develop resistance to fungicides more rapidly. How this translates into orchard practice is that we need to be careful in our use of sterol-inhibiting fungicides. Orchard practices that discourage the development of fungicide-resistant pathogen strains should be followed.

We have studied the effect of temperature, time, and relative humidity (RH) on the germination of conidia on leaf surfaces. The optimum temperature for conidial germination is about 70°F, with the highest numbers of conidia germinating at humidities above 90 percent. However, the conidia will germinate in lower numbers at humidities as low as 40 percent. Our research also indicates that the conidia require between 16 and 24 hours of incubation at disease-conductive temperatures and humidities to reach optimal germination

FIGURE 3

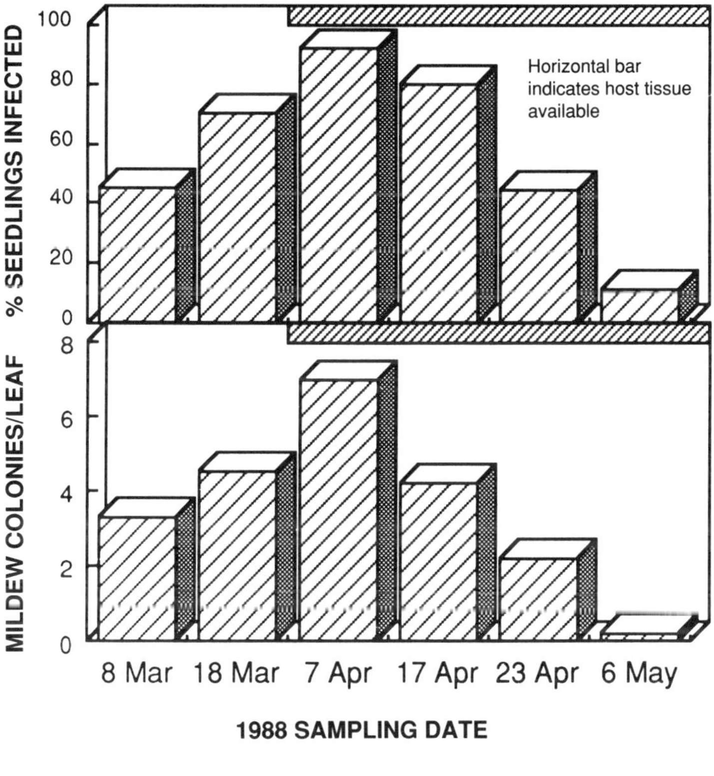

rates; germination after eight hours was generally very low. This is important in that it shows that an epidemic of powdery mildew requires more than our normal late spring/early summer night temperature/humidity combinations. Periods of rain are needed to provide longer periods of high humidity that favor the fungus. Experiments have also shown that the conidia will germinate readily on green fruit. On green fruit, germination required 16- to 24-hour incubation periods at high humidity. Germination on ripening fruit was suppressed by increasing sugar concentrations. At sugar concentrations more than 15 percent, germination ceased. This indicates that fruit is more likely to become infected when green or when sugar concentrations are lower; infection is highly unlikely when fruit sugar contents are high.

At this time, sulfur and Rubigan are the only fungicides that we have for control of the disease. Sulfur is purely a protectant fungicide and must be on the leaves or fruit before the spores of the fungus arrive. Spraying sulfur after the actual infection takes place is not going to control the disease. Rubigan is a more effective compound and is the best choice during wet years when disease pressure is high and/or in orchards with a mildew history. It cannot be overstated that keeping mildew off foliage during May and June is the key to keeping it off fruit. It is essential to follow a good fungicide spray program. Begin applying fungicides at (the latest) shuck fall. If we have a significant rain event between bud burst and shuck fall, a Rubigan application as soon as things dry is probably a good idea.

4

SPRINKLER ROT OF APPLE AND PEAR

GARY G. GROVE
Washington State University
Tree Fruit Research &
Extension Center
Wenatchee, Washington

SPRINKLER ROT IS PRIMARILY A FRUIT ROT OF PEAR, but can occur less commonly on apples. It is caused by the same fungus that causes collar rot of apple—*Phytophthora cactorum*. The disease is most problematic in orchards with overtree irrigation, but can occur in orchards with undertree sprinklers if the angle of the sprinklers is sufficiently high to wet the fruit on the lower portions of the tree.

On immature pears, the disease can sometimes be confused with fire blight. Sprinkler rot lesions are typically light brown in color and softer (fire blight on immature fruit is dark brown and hard); infected fruit frequently have a pungent, phenolic odor. An accurate diagnosis can sometimes be made by placing infected fruit in a moist incubation chamber and incubating at about 70°F for three to four days. After that time period, fruit with sprinkler rot will have a very obvious white cottony growth.

The fungus survives winter in the form of a survival propagule called an oospore. Oospores are in the soil at the canal bottoms. When the canals are flooded, the oospores germinate, producing a sporangium. The sporangia produce zoospores that swim through the water and probably infect roots of weed hosts growing along the canal. Sporangia of the host are probably produced on submerged host roots; zoospores (the infective propagule of the fungus) are produced within the sporangia and liberated into the irrigation water. When the water is shut off in the fall, the infected host roots die and the fungus within probably enters its sexual (survival) stage and produces oospores to perennate the fungus until the water is turned on the following year.

The use of overtree irrigation with contaminated water is actually providing perfect conditions (inoculation and incubation) for disease development. We are delivering the pathogen to the host and providing the wetness conditions necessary for fruit penetration. Our sprinkler rot research program has been designed to answer several questions: 1) when is the fungus present in irrigation water, 2) during what stages of fruit development are fruit susceptible, and 3) what temperature and wetness conditions are necessary for fruit infection. By periodically baiting irrigation canals with pear fruits, we have determined that the fungus is present from mid May until late September.

A three-year study of the temperature and wetness requirements for infection of apples and pears has been completed. By inoculating fruit with zoospores and subjecting them to a variety of temperature: wetness period regimens, it was found that as little as one hour of wetness was required for infection between 62 to 85°F. Progressively longer wetness periods are required as temperatures decrease from 62°F, but the times required are still well short of those provided by our standard irrigation set. As temperatures increase up to 85°F, infection requires much shorter wetness periods. Although our data indicate that we cannot manage this disease by reducing the length of our irrigation sets, we have not found the high temperature infection cut-off. In the process of the study, it was demonstrated that pears are susceptible from about one month after petal fall until harvest; susceptibility did not increase or decrease significantly over that time period. The susceptibility of apples increased significantly during the 10 days preceding harvest.

In overtree blocks, the only means of control available is the use of copper in irrigation water. It needs to be metered into the water at one part per million. In d'Anjou blocks, it is important for the prevention of fruit russetting that this concentration not be exceeded.

Simply placing copper in a burlap bag and throwing it into the irrigation water is not advisable; an injection system is the only means for treating water with any degree of accuracy. If the disease is a problem in undertree blocks, lower the sprinkler water angle so that wetting of fruit surfaces is avoided.

5

FIRE BLIGHT PREDICTION AND CONTROL

SHERMAN THOMSON
Utah State University
Department of Biology
Logan, Utah

THE FIRE BLIGHT BACTERIUM, *ERWINIA AMYLOVORA*, apparently evolved on indigenous American plants such as hawthorn and native crab apples. Damage to these native plants was probably minimal and usually caused only the loss of a few flower clusters. The bacteria and hosts had evolved to a stage where they coexisted, and, indeed, the disease may have been been beneficial to the pomaceous hosts by accomplishing some random "pruning and thinning."

When the colonists arrived in North America in the early 1600's, they brought with them seeds and plants of pear, apple, and quince from England and France. These introduced hosts were far more susceptible to fire blight and major outbreaks of the disease occurred. By about 1900, fire blight was so severe in the eastern United States that the planting of the European pear was discontinued as a commercial orchard tree. It took fire blight 135 years to move from the eastern United States to the West Coast of North America, generally following the settlement of the country. Apples are generally more resistant to fire blight, and therefore the apple industry in the northeastern United States expanded while pear planting declined.

Fire blight continues to spread throughout the world and is considered a major concern wherever pome fruits are grown. Almost each year, a new country or area reports the first occurrence of fire blight. It is now found in 19 countries on four continents—all of North America, much of Central America, Europe, the Near East, and New Zealand.

DISEASE CYCLE

The primary overwintering sites of the fire blight organisms are the cankers originating from the previous season's infections. These cankers often exude bacterial ooze in the spring as temperatures warm. This ooze is usually produced from indeterminate cankers without a definite margin and is composed of viable bacteria in a matrix of polysaccharide. It is attractive to many insects, which become contaminated with the bacteria during feeding. These insects are also attracted to flowers while seeking pollen or nectar. In these visits, the insects inadvertently deposit the pathogen on flower surfaces. Wind and splashing rains are also responsible for spreading bacteria in the orchard.

The bacterium enters tissues through natural openings such as nectaries, hydathodes, and lenticels, or through wounds caused by insects. Meteorological events such as hail or driving rain create minute injuries, allowing entry of the pathogen. Flowers are the most susceptible, but infections of succulent shoot tips are also common.

Epiphytic bacteria develop on apparently healthy pear flowers, fruits, and leaves in response to specific environmental conditions. Populations are usually detected in the western states on healthy flowers only after the daily mean temperature exceeds 60°F. Epiphytic populations are often present in healthy flowers seven to 20 days before fire blight symptoms are noted. In many cases, these populations exceed 100,000 cells per healthy flower. The occurrence of fire blight outbreaks can be directly related to the presence of epiphytic populations of bacteria on host flowers.

Epiphytic populations generally have not been detected in orchards in midwestern or eastern states prior to fire blight outbreaks. It has been speculated that because the environmental conditions are more regularly conducive for fire blight with frequent rains and heavy dews, that epiphytic populations do not develop extensively.

Epiphytic populations of bacteria are present almost exclusively on the stigmatic surfaces of the pistils of flowers of pear, apple, and ornamental hosts. Rain provides a mechanism for movement of the epiphytic bacteria on the stigmatic surface to other flower parts, resulting in infections. Fire blight outbreaks almost always occur with a nearly simultaneous expression of symptoms on flowers or shoot tips. Random distribution by insects would not result in sudden outbreak if infection occurred as a result of the mere deposition of the bacteria on the flower. Rain dispersal results in localized outbreaks near oozing holdover cankers or the distribution of infections in an inverted cone-shaped area.

These results explain why epidemics occur so rapidly following a rain storm. Insects or rain are responsible for inoculating flowers with low populations of bacteria, which at the time may not initiate an infection. The stigma provides the proper environment and nutrients for the populations to increase to levels of 100,000 to 10,000,000, and when rain or heavy dew occurs, these high populations are transferred en masse to the hypanthium, where infections occur nearly simultaneously in the orchard. However, symptoms may not be apparent for several days.

Rain is often associated with fire blight outbreaks but is not necessary. High relative humidity or heavy dew are sufficient to promote infection.

Inoculum for secondary infection originates from droplets of ooze produced on infected flowers or terminals. This ooze, like that on overwintering cankers, is attractive to insects, which quickly spread the inoculum through the orchard. Rain or heavy dew randomly disseminates the bacteria to potential infection sites on flowers or terminals. Strong winds associated with thunderstorms and hailstorms are ideal for initiating epidemic blight, because these conditions generate the wounds necessary for infection as well as transfer the bacteria.

Insects with piercing or sucking mouth parts are instrumental in causing outbreaks of shoot blight. The pear psylla and tarnished plant bug have been shown to initiate fire blight infections during normal feeding activity on flower clusters and leaf axils. White apple leafhopper and aphids are also likely candidates as vectors.

The bacteria survive the winter in cankers which develop on all hosts. Large cankers are easy to locate and eradicate. However, there are many small cankers which are formed on twigs that go unnoticed. The inconspicuous cankers may be responsible for major sources of inoculum in the spring. Reducing the number of overwintering cankers is one of the major control strategies that every grower should utilize. However, even when growers are extremely thorough in pruning, fire blight can still develop under conducive environments, because massive amounts of bacteria can be produced from a single canker.

The influence of environment is undoubtedly the most important factor in fire blight epidemics, but is also the aspect of the disease we are least able to predict or control. These deficiencies in knowledge seem more acute when fire blight devastates an assumed resistant cultivar or attacks trees in a geographical area not normally plagued with the disease. Despite the numerous studies on the epidemiology of fire blight and improved bactericides, the disease is still not always satisfactorily controlled.

CONTROL

Fire blight is an erratic but potentially devastating disease that has prompted pear and apple producers in the United States to spare little in their efforts to control the disease. Growers rely heavily on sprays during the bloom period to protect their orchards. Despite these treatments, fire blight can still be a threatening disease that cannot be controlled by a few simple measures. Satisfactory control can be obtained only by utilizing an integrated management program that includes many cultural and chemical methods.

Cultural Control

Fire blight will be less of a problem and fruit of better quality will result when orchards are planted on well-drained sites with light soils and with a slightly alkaline soil (pH of 6.0 to 6.5). Studies have shown that trees grown on heavy or poorly drained soils sustain more blight, and the damage from blight is more extensive than on trees grown on favorable sites. This is partly due to the slower release and delayed availability of nitrate nitrogen on poorly drained sites. Nitrate nitrogen is the form used by the plant, whereas the ammonium ion must be converted to the nitrate form before being used. Under cool soil temperatures, the microorganisms that cause nitrification (conversion of ammonium ion to nitrate) are not active, and therefore, the trees grow slowly. As soil temperatures increase, more nitrate becomes available and trees respond with vigorous and susceptible growth, at a time when ambient air temperatures are more conducive to fire blight activity.

Irrigation procedures also influence the incidence of fire blight. Sprinkler irrigation which wets the foliage or blossoms spreads the organism and provides the moisture necessary for initiating infections. Sprinklers with low angles should be used to avoid wetting the foliage. Overhead irrigation should not be used where blight is a problem.

Fertilization with nitrogen or cultivation (which makes more nitrogen available to trees) results in more fire blight infections and greater lesion length. Most growers are aware of the effect of nitrogen on increased succulence and subsequent increases in fire blight susceptibility, but some growers withhold nitrogen to the extent that tree growth and yields are seriously affected.

Nutrient application should be based on soil and leaf analyses and on levels of growth observed in previous years. Excess nitrogen or organic sources of nitrogen, such as animal manures, should not be applied to orchards where fire blight is likely to be a problem.

Fruiting spurs and suckers on the trunk or major scaffold branches should be pruned out to avoid infections which might kill the tree or cause serious damage to the tree form or structure.

The removal of overwintering cankers during dormant pruning operations is one of the most important cultural practices to reduce subsequent blight in an orchard. Most cankers can be easily detected and removed during the normal dormant pruning practices. However, it is difficult to completely eradicate all cankers in an orchard. Cankers with indeterminate margins and those on small twigs frequently go undetected. These cankers provide sufficient inoculum in an orchard for a serious epidemic to develop.

The fire blight pathogen is present in tissues in advance of externally visible symptoms. Therefore, cuts should be made at least six to 12 inches ahead of the visible lesion. Failure to do so may result in the extension of the infection or contamination of the pruning tools. Highly susceptible varieties and succulent trees may require more aggressive pruning, in excess of one foot into healthy wood. Pruning during the dormant season and on hot, dry days requires no sterilization of tools. However, when pruning active infections during moderate summer weather, it is advisable to dip tools in 10 percent bleach (sodium hypochlorite) or 70 percent methanol.

Insects are very efficient vectors of *E. amylovora* from ooze on overwintering cankers to flowers, as well as transfer of the pathogen from flower to flower. However, the activity of these insect vectors is so closely intertwined with the essential pollination activity that early season vector control during bloom is not recommended in fruit crops.

Aphids, leafhoppers, psylla, tarnished plant bugs, and other sucking insects are considered important vectors responsible for shoot blight. The control of sucking insects in the postbloom period is recommended, partly to reduce the actual damage caused by the insects and partly to reduce the potential for initiating fire blight infections on vegetative shoots. This approach seems to be particularly important if an adjacent field of alfalfa or another crop is harvested, creating a mass migration of insects.

Chemical Control

Orchards with a previous history of fire blight and highly susceptible cultivars should be sprayed with an 8-8-100 Bordeaux mixture, plus one percent dormant oil at silver- to green-tip stages. This concentrated copper application apparently reduces the production of inoculum from overwintering cankers. Other copper compounds are also effective.

The protection of flowers by bactericide sprays in most years is usually an effective method of controlling fire blight, even in areas where conditions and inoculum are conducive for disease. However, the level of control obtained by the currently available bactericides is less than ideal. None of the registered products in the United States have eradicating or significant systemic properties.

The timing or prediction of when to apply bactericides is a critical technique. Our studies have shown that flowers can be protected by initiating the first bloom sprays in the spring as soon as the daily mean temperature exceeds 60°F. Thereafter, applications should be repeated every three to five days. The daily risk is determined by temperatures and precipitation. On high risk days, a bactericide should be applied within 24 hours on each row. Moderate risk days may require applications only in highly susceptible orchards.

Bordeaux mixture or other fixed coppers provide satisfactory results in most cases, but when conditions are highly conducive for fire blight, they are not very effective. Unfortunately, copper compounds may cause russetting of light-colored fruit varieties, especially pears.

The best material available for fire blight control is streptomycin. However, to be effective it must be applied at the proper times, since it is not an eradicant and only effective on open blossoms for two to five days. Therefore, applications must be repeated every three to five days to protect newly opened blossoms and provide a protective residue.

The development of strains of *E. amylovora* resistant to streptomycin in many West Coast fruit orchards has greatly complicated control measures. The application of streptomycin in orchards with resistant strains is virtually useless and results in nothing but a loss of money.

Oxytetracycline is an alternative antibiotic that is also registered for fire blight control on apples but is generally not as effective as streptomycin. In areas where streptomycin resistance is not a problem, growers should consider tank-mixing the two antibiotics or alternating bactericides to prevent the selection of resistant strains and preserve the usefulness of these compounds.

Biological Control

Biological control of fire blight with antagonistic bacteria has been demonstrated experimentally in many areas, and as long ago as 1930. In some cases, the use of antagonistic bacteria has been as good as streptomycin, but results have been inconsistent. High populations of the beneficial bacteria are sprayed on orchards in full bloom to protect the flowers. The antagonistic bacteria apparently colonize the stigmatic surface of flowers and prevent the development of the fire blight bacteria.

We are investigating the use of honeybees as vectors to disseminate antagonistic bacteria to flowers to protect against fire blight infections. The bees acquire the beneficial bacteria when exiting the hive and deposit them on the stigmata when visiting flowers.

Fire blight is still a major threat to pome fruit producers and the ornamental industry worldwide, despite the large accumulation of knowledge that is available regarding the epidemiology and control of the disease. There are some significant gaps in our understanding of the disease, and only through continued research efforts will the threat of fire blight be reduced. Fire blight has become an important international disease, and cooperative programs are instrumental in its control.

6

FIRE BLIGHT PREDICTION AND CONTROL

PAUL W. STEINER
Department of Botany
The University of Maryland
College Park, Maryland

FIRE BLIGHT, CAUSED BY THE BACTERIUM *ERWINIA amylovora,* is one of the most sporadic and destructive diseases affecting commercial apple and pear production in the United States. While serious losses from some diseases can be avoided by improving the control tactics after early symptoms appear, this is not true with fire blight. Inoculum from even a few fire blight cankers is dispersed widely and increases epiphytically well in advance of when infections take place, if they occur at all. An infection event can occur in a matter of hours and, by the time symptoms appear a week to a month later, the disease can be so well established that little can be done to stop the inevitable progress of the epidemic. For this reason, the only practical control strategy is to take effective action before infections occur.

Our best control programs involve thorough pruning and early copper sprays to reduce primary inoculum, intensive spraying to protect the very susceptible blossoms, and balanced fertility to avoid excessively succulent vegetative growth. This approach, while often adequate, is costly, seldom provides complete control, and, sometimes, fails with serious losses. Such inconsistencies in control all point to a basic problem involving the precise timing of treatments. These faults can be corrected only by gaining a more detailed understanding of the infection process and the factors that control the progress of epidemics. A new fire blight forecasting system developed at The University of Maryland, called MARYBLYT, offers some answers but, at the same time, raises many new questions. Nevertheless, progress is being made, and the focus of this discussion is to summarize our current knowledge and to suggest ways in which recommendations for controlling fire blight might be improved.

MULTIPLE FORMS OF FIRE BLIGHT

There are at least five distinct types of infections associated with fire blight symptoms, each developing from a different source of inoculum and in response to different conditions. Blossom blight symptoms result from direct infections of open flowers, with subsequent invasions of fruiting spurs and supporting limbs. Canker blight involves renewed infectious activity by the pathogen at the margins of indeterminate cankers, with subsequent invasions of the cankered limb and nearby shoots. Shoot blight symptoms result from direct infections of vegetative shoot tips, with subsequent invasion of supporting limbs. Trauma blight symptoms develop on many different tissues and are associated with infections following injuries caused by hail, late frosts, or wind. Rootstock blight involves the secondary invasion and subsequent infection of susceptible rootstocks (or interstems) by the pathogen, following a latent residency in healthy scion tissues. It is important to understand the nature of each of these types to predict fire blight infection events and to focus specific control measures more precisely.

Blossom Blight

There are four strict requirements for blossom infections, all of which must occur in the following sequence: 1) flowers open with stigmas and petals intact; 2) development of a minimum inoculum potential with the accumulation of at least 198 degree-hours greater than 65°F from full pink (= first bloom); 3) a wetting event occurring as dew or greater than or equal to 0.01 inch rain or greater than or equal to 0.10 inch of rain the previous day; and 4) an average daily temperature greater than 60°F.

BLOSSOM FACTORS. Figure 1 shows that the proportion of flower buds available as infection courts varies during the bloom period and can be predicted on the basis of cumulative degree-days greater than 40°F from green tip. An individual flower remains open for only about 80 cumulative degree-days greater than 40°F, or about four days with an average temperature of 60°F. Because single flower infections usually invade and destroy the fruiting spur supporting five blossom buds, the loss risk of any infection event must be based on the proportion of spurs with one or more open blooms (i.e., > 20% bloom or about 1 open flower/spur). Cultivars like Rome in Maryland often have a secondary flowering habit that lengthens the risk period. This is also important from an overall loss standpoint because blossom infections occurring during the strong vegetative growth phase of the tree tend to extend further into the supporting limb.

FIGURE 1

MARYBLYT apple bloom models for Jonathan, York, and Rome cultivars in Maryland. Note varietal differences in the length of the high risk period.

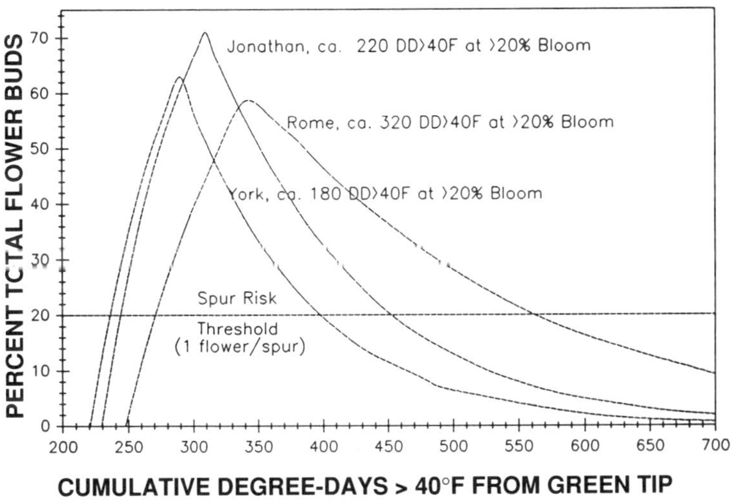

INOCULUM POTENTIAL. Despite the fact that MARYBLYT was developed for use on apples in Maryland, data from California on pear blossom colonization, with some adjustments in calculation, have worked very well in nearly all locations where the model has been tested in the United States. Figure 2 shows this relationship is logarithmic, which means that the epiphytic inoculum potential increases very rapidly. It is assumed that the bacteria multiply and selectively colonize the stigma surfaces of apple flowers and that cumulative degree-hours greater than 65°F is an indirect measure of inoculum potential. The cumulative degree-hour total, accrued from full pink, is reduced to zero in response to any day with a minimum temperature less than or equal to 32°F or three consecutive days with a maximum temperature less than 65°F. Proportional reductions for one or two consecutive days are also made until greater than or equal to 400 cumulative degree-hours. This does not mean that cool weather kills the bacteria. On the contrary, the negative adjustments are needed to reflect the effect of several factors on the overall epiphytic inoculum potential: 1) loss of colonized flowers through normal maturation; 2) reduced rate of newly opened flowers being colonized; 3) reduced bee activity; and 4) loss of bacteria already acquired by the bee population.

FIGURE 2

Relationship between percent open Bartlett pear flowers colonized by *E. amylovora* and cumulative degree-hours > 65°F.

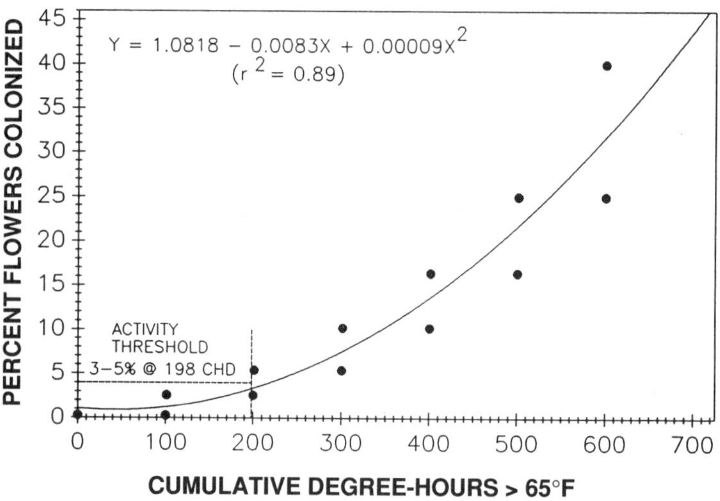

The combination of data from Figures 1 and 2 can be used to develop risk estimates for any given infection event based on the cumulative degree-hours greater than 65°F and the proportion of flowers open. Because the stigma surfaces are colonized selectively, the first requirement for the successful establishment of an inoculum potential is for a single bacteria to be deposited on one of five stigmas in one of five flowers per spur. Once the first few open flowers are colonized, the activity of pollinating insects appears to be the dominant factor in the rapid dispersal of inoculum throughout the planting.

WETTING FACTOR. Once flower stigmas are colonized, a wetting event is required to move the bacteria into the nectary in the base of the flower, where entry into the nectarthodes and infection occurs. Since the bacteria are highly motile and positively chemotaxic for any of several simple compounds in apple nectar, the minimum requirement for allowing their inoculation from the stigmas to the nectarthodes is simply a continuous film of water between the two points. In this respect, a heavy dew can be more effective in triggering a severe infection event than a light rain. Be aware, too, that when the epiphytic inoculum potential is high and temperatures are suitable, an infection period can be triggered by spray applications or overhead irrigation during bloom.

FIGURE 3

Estimated loss of apple fruit spurs due to primary blossom infections by *E. amylovora* at various stages of flowering.

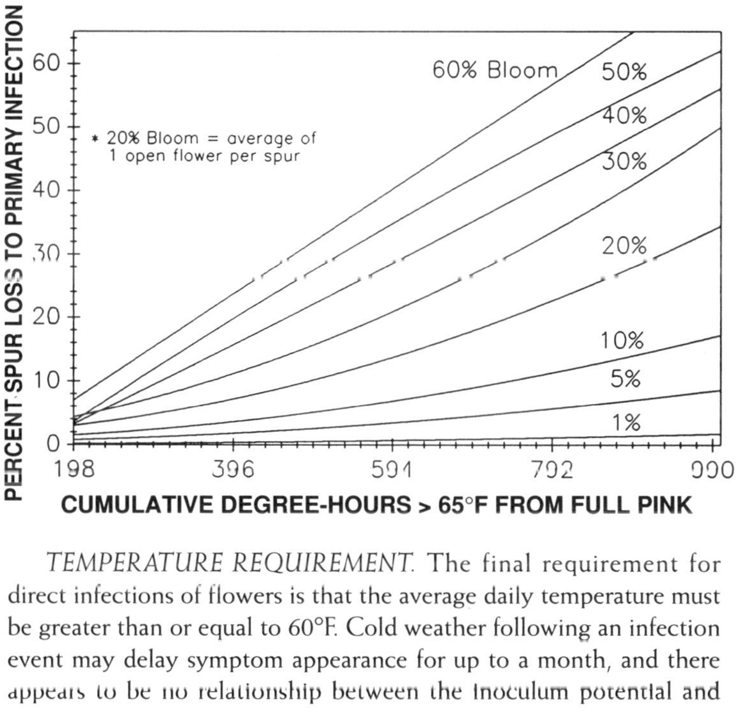

TEMPERATURE REQUIREMENT. The final requirement for direct infections of flowers is that the average daily temperature must be greater than or equal to 60°F. Cold weather following an infection event may delay symptom appearance for up to a month, and there appears to be no relationship between the inoculum potential and the rate of symptom development under field conditions. Thus, the required temperature may be related more to rate of bacterial motility than to multiplication of the bacteria, since motility is reduced at temperatures less than 65°F or greater than 86°F.

INCIDENCE AND SEVERITY. The above conditions represent only the minimum requirements for infection to occur. Whether a given infection event will result in mild, moderate, or severe damage depends on the degree to which any one or more of these thresholds is exceeded. Following infection, early symptom appearance can be predicted and confirmed by close observation with the accumulation of at least 90 degree-days greater than 55°F. This infection-to-symptom interval has been an extremely reliable predictive index for the MARYBLYT program and appears to be the same for all other types of infections, with the possible exception of rootstock blight.

MAKING TREATMENT DECISIONS. Table 1 describes the system used in the MARYBLYT program for reporting the risk of blossom infections. When the risk becomes moderate to high, users should trigger the program's prediction option and enter the forecast weather data for the next one to three days to determine the likelihood of an infection event. The need for subsequent treatments should be reassessed not more than four days following the first treatment.

TABLE 1

MARYBLYT system for reporting the minimum risk of blossom infections occurring.

Risk of infection	Bloom	Requirements for infection CDH>65F	Wetting	Ave>60F
None	-	na	na	na
Low	+	-	-	-
Moderate	+	+	-	-
or	+	-	+	-
or	+	-	-	+
High	+	+	+	-
or	+	+	-	+
or	+	-	+	+
Infection*	+	+	+	+

* *First symptoms in 90 degree-days greater than 55°F.*

Under Maryland conditions, we have obtained excellent control of blossom blight using streptomycin 17 percent SP at 0.75 to 1.0 pound formulation per 100 gallons of tank mix (150-200 ppm) with a delivery rate of 0.09 to 0.11 gallons of spray per 1,000 cubic feet of tree row volume (40-60 gpa, depending on tree size and planting density). An activator adjuvant such as Regulaid™ (1 part/100 gallons) is always recommended. Treatments are recommended via a telephone alert system for the day before or the day of an anticipated infection event. Usually only one or two sprays are needed, with the second spray often aimed at delaying an anticipated outbreak of shoot blight. At the 40 gallon per acre delivery rate, these recommendations represent approximately 0.30 to 0.40 pounds per acre per treatment, compared with the standard recommendation elsewhere for three to four routine sprays using 1.5 to 2.0 pounds per acre.

Canker Blight

Unlike the sporadic appearance of blossom blight, where overwintering cankers are present, canker blight occurs regularly every year. Figure 4 shows the development of canker and shoot blight can be predicted reliably on the basis of cumulative degree-days greater than 55°F better than by using calendar days. The earliest symptom of canker blight occurs in the inner bark at indeterminate margins of overwintering cankers with accumulation of at least 130 degree-days greater than 55°F after green tip. This canker margin

FIGURE 4

Sequential appearance and prediction of canker and shoot blight development relative to cumulative degree-days > 55°F from apple green tip.

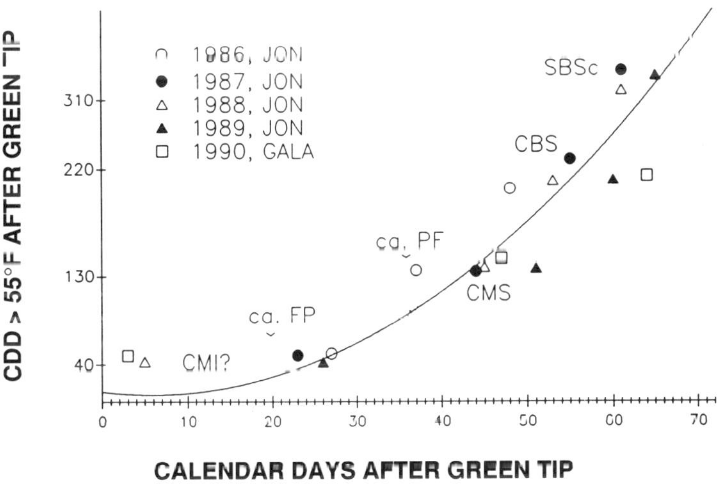

symptom is a water soaking with diffuse brown discoloration in the living bark tissue surrounding a canker. Classic canker blight symptoms show wilt, necrosis, and death of nearby water sprouts or branch tips, which is first evident 99 degree-days greater than 55°F after canker margin symptom (i.e., 220 CDD > 55°F after green tip).

The repeated visual confirmation of the 90 degree-day interval between canker margin symptom and canker blight symptom every year since 1986, suggests that at 40 cumulative degree-days greater than 55°F after green tip, canker margin infections are most likely to begin (i.e., 90 degree-days before canker margin symptom). Experiments to determine whether the first appearance of ooze on bark surfaces at canker margins is coincident with canker margin infection are underway. In seasons when bud break is initiated early in response to abnormally warm temperatures, as in 1988 and 1990, canker margin infection is predicted to have occurred shortly after green tip. Canker margin infection appears to be triggered by a combination of ambient temperature and cumulative heat units (DD) affecting bacterial multiplication and changes in the mobilization of stored carbohydrates relative to vegetative bud development. Canker blight symptom usually occurs one to two weeks after petal fall on apples and, along with any spurs showing symptoms of blossom blight, provides the primary source of inoculum for the shoot blight phase that follows.

CANKER BLIGHT CONTROL. There are no recommended treatments for canker blight control other than thorough pruning on an annual basis during the dormant season. Based on the data in Figure 4, pruning after canker margin infection may well serve to spread the pathogen elsewhere in the orchard.

Shoot Blight

Shoot blight is largely a secondary event in that the sources of inoculum are tissues showing symptoms of either blossom or canker blight in the vicinity (i.e., not necessarily in the monitored block). Infections occur as the result of direct inoculation of shoot tips and appear to be accomplished largely by insect vectors with sucking mouthparts. New shoot blight strikes often occur at lower levels in the tree under existing blossom or canker blight symptoms above following a rain.

In Maryland, the earliest potential insect vectors for *E. amylovora* shoot infections are adults of the white apple leafhopper (first trapped at about 650 CCD > 40°F after green tip). Alate green apple aphids follow at about 900-1000 CDD greater than 40°F after green tip. Early symptoms (slight tip wilt, sometimes a blackening of the mid rib of the third leaf from the tip) or ooze droplets appear 90 degree-days greater than 55°F after all of the above conditions are met. If blossom or canker blight symptoms develop before vectors become available, then symptoms appear to be delayed until 90 degree-days after vectors do appear. If the monitored orchard has been generally "blight free" and the orchard is fairly isolated, shoot blight may occur only very late in the season, causing little overall damage and is often restricted to the outside margins of the orchard. We cannot predict this latter incidence.

SHOOT BLIGHT CONTROL. Treatments for shoot blight control in the past have never proven very effective, so they are often specifically not recommended. Most programs, however, begin only after symptoms first appear, when many other secondary infections have already been established, with symptoms inevitably showing later in spite of streptomycin treatments. An early application of streptomycin after petal fall, but before the first anticipated shoot blight infections occur, does appear to reduce the amount of shoot blight by delaying the epidemic long enough for normal shoot maturity to confer some greater degree of resistance. Our best success has been with growers applying a streptomycin plus activator plus vector insecticide tank mix just prior to the first anticipated shoot blight infections, followed by close monitoring and treatments for insect vectors until terminal bud set is complete (shoot tip leaves are fully expanded and more resistant).

Pruning out infected shoots as they appear has not been a reliable practice in Maryland and, in some instances, may increase the potential for additional infections. If this is attempted, cuts must be made very early, before necrotic tissue develops, and all cutting operations should cease if the epidemic seems to continue. Once necrosis appears, the bacteria can be readily isolated from apparently healthy wood several feet in advance of the symptom margin, a limb length most growers are reluctant to sacrifice. Excessive pruning stimulates near shoot growth, which only extends the period of susceptibility. Under some conditions, the natural resistance conferred by maturing tissues may be reduced by the simple wounding involved in making the cuts.

Trauma Blight

Trauma blight symptoms result from infections that develop in association with injuries caused by late frosts of less than or equal to 28°F, hail, or high winds when the epiphytic inoculum potential is high (usually > 500 CDH > 65°F). Infections incited by frost can breech the defense mechanisms in normally resistant cultivars like

Red Delicious. Hailstorms, high winds, and late frosts are difficult to predict in advance. When they do occur, any treatment decision must be made immediately, since those made after symptoms appear 90 degree-days greater than 55°F later are not likely to be effective. We have no data on how cost-effective specific streptomycin treatments might be in such situations.

Rootstock Blight

The growing popularity of M.26 and M.9 dwarfing rootstocks presents a serious risk where the scion cultivar is also susceptible to fire blight. These rootstocks are extremely susceptible to fire blight and can be killed in a single season after the scion sustains only a few, apparently minor, shoot or blossom infections. Observations and isolations in several Maryland locations in 1990 suggest that the bacteria move quickly in the intercellular spaces of the bark cortex in mature, healthy limbs and trunks, invading the rootstock by late June to early July. Some cankering may be initiated in the rootstock then, and the pathogen can be easily isolated there throughout July and August. Symptoms of tree decline and early red fall color may not be apparent until late August to early September and, for this reason, are sometimes mistaken for collar rot (caused by the fungus *Phytophthora cactorum*) rather than fire blight. In 1990, we found several incidents of rootstock blight and subsequent tree death in five- to nine-year-old orchards of Gala/M.9 or M.26 and Jonee or Jonathan/M.26, with some losses as high as 10 to 30 percent. In the future, M.7 rootstocks may be a better choice for use with fire blight-sensitive scion cultivars.

Summary

Fire blight epidemics within a given orchard in any given season can develop as any one or a combination of three major phases, identified here as blossom, canker, and shoot blight. Trauma blight and rootstock blight are unusual, but not uncommon, and present additional risks for very high tree losses. Each of these blight types can be distinguished on the basis of symptoms, the tissues affected, sources of inoculum, and by the conditions required for infection. This information, coupled with the ability to predict the occurrence and progress of each phase reliably and accurately, offers many opportunities for developing more efficient and cost-effective disease management strategies and tactics. We anticipate that the validation and documentation supporting the MARYBLYT program will be complete in 1991 and that the program will be available to growers in 1992.

REFERENCES

Bayot, R. G. and S. M. Reis, 1986. Role of motility in apple blossom infection by *Erwinia amylovora* and studies of fire blight control with attractant and repellent compounds. *Phytopathology* 76:441-445.

Lightner, G. and P. W. Steiner, 1990. Computerization of blossom blight prediction model. *Acta Horticulturae* 273:159-173.

Steiner, P. W., 1990. Predicting apple blossom infections by *Erwinia amylovora* using the Maryblyt model. *Acta Horticulturae* 273:139-148.

Steiner, P. W., 1990. Predicting canker, shoot, and trauma blight phases of apple fire blight epidemics using the Maryblyt program. *Acta Horticulturae* 273:149-158.

Thomson, S. V., 1986. The role of the stigma in fire blight infections. *Phytopathology* 76:476-482.

Zoller, B. G. and J. Sisevich, 1979. Blossom populations of *Erwinia amylovora* in pear orchards vs. accumulated degree-hours over 18.3°C (65°F), 1972-1976. *Phytopathology* 69:1050.

Part 2

Fundamental Principles of Integrated Pest Management

7

USE OF PHENOLOGY MODELS FOR PEST MANAGEMENT

VINCENT P. JONES
Department of Entomology
University of Hawaii
Honolulu, Hawaii

PHENOLOGY MODELS ARE ONE OF THE MAJOR components of pest management programs which allow researchers to decide on optimal timing of pesticide applications in a system with multiple pests present. These models do not operate on a calendar day basis, but instead, operate on a heat unit scale, which insects respond to in a very predictable manner. The model is generally started from some easily detected event, such as first flight of codling moth, and used to predict something that is important, but not easily sampled (e.g., 3% codling moth egg hatch). In this chapter, I will discuss the basis of phenological models, how they are developed, which ones are available, and give examples of proper and improper use of them.

HOW AND WHY THEY WORK

Metabolic rates of animals are related to the temperatures at which those metabolic reactions occur. In biological (enzymatic) reactions, there is generally a temperature at which the reaction goes fastest. If you raise the temperature above the optimal temperature, the precise fit of the enzyme on its substrait is disrupted, and the reaction rate rapidly decreases. If the temperature remains high, the enzyme is generally permanently disrupted, and death occurs. If you decrease the temperature below the optimal temperature, the rate of the reaction drops rapidly until it stops, but death is rarely the result unless freezing and ice crystal formation occurs in the animal's cells. In warm-blooded animals, body core temperatures rarely vary more than a few degrees, and the animal's enzymes tend to be operating at peak efficiency near the animal's normal body temperature.

This means the rate of enzymatic reactions is fairly constant and that development of warm-blooded animals can therefore be readily predicted by calendar time. However, for invertebrates like insects and mites, there are no physiological mechanisms which regulate body temperatures closely. This means that, in general, their body temperatures are directly related to ambient temperatures and therefore, their development can best be predicted from heat accumulations and not calendar time.

Formally, the idea of heat accumulation being used to predict development is known as the law of constant thermal summation. Basically, it says the total number of heat units required to complete a given physiological process is constant (within the upper and lower limits described above). The rate at which the heat units accumulate, whether quickly or slowly, is immaterial. The heat unit scale is often called a physiological time scale.

A good analogy to the law of constant thermal summation would be the filling of a gallon container. It doesn't matter how fast or how slow you fill the container, it still takes a gallon to fill it. If you consider the gallon container to be a physiological process, like the development through the egg stage, the amount of heat required is the same whether the heat is accumulated quickly or slowly (within certain bounds).

The temperature limits on physiological reactions are termed the upper and lower developmental thresholds. When the temperatures drop below the lower threshold, no development occurs; conversely, when the temperatures rise above a certain temperature, at first development stops, and if temperatures continue to rise, thermal death occurs. To measure the number of heat units accumulated, degree-days are used. A degree-day is defined as the heat experienced by the organism when the temperature is one degree above the lower threshold for development for a period of 24 hours.

Most models use a sine-wave to approximate the daily temperature cycle from night to day. The lower threshold is frequently called

K1, from the terminology developed by the individuals who developed the sine wave approximation. The upper threshold can have at least two forms (*see Figure 1*): 1) a horizontal cutoff, where heat unit accumulations do not rise above a maximum rate at the upper threshold for development (K2); and 2) a vertical cutoff (K3), where once the upper threshold is surpassed, no additional heat units are accumulated. The heat accumulated is calculated as the area under the curve within the upper and lower thresholds (dark

FIGURE 1

Heat unit accumulation using (A) vertical cutoff and (B) horizontal cutoff. Dark areas are the heat accumulation for each type.

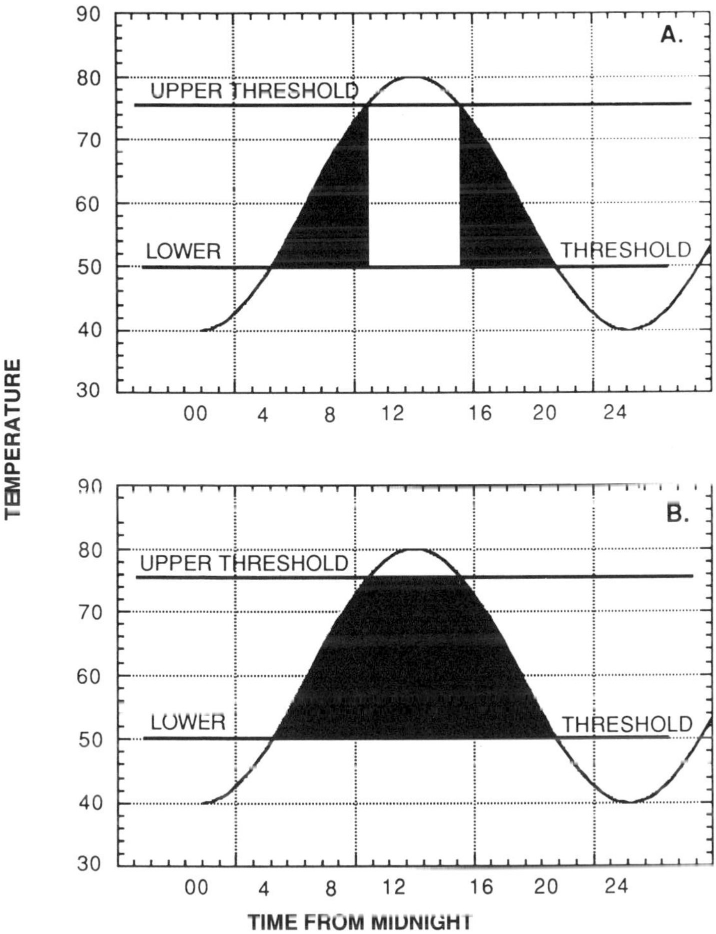

areas in Fig. 1 are the heat units accumulated). The first condition does not appear biologically reasonable in light of this explanation, but it does seem to provide better predictions in certain circumstances. A table is often generated which allows the end-user to merely look up the maximum and minimum temperatures for a day and obtain the number of heat units accumulated during that period for a particular insect. These daily heat units are accumulated from the biofix and are used to predict when the critical event occurs.

Biophenometers are also used to accumulate heat units. They are programmed with the desired upper and lower thresholds for development and placed in the orchard at a particular time. The grower can then just go to the weather shelter and, at the press of a button, get the accumulated number of heat units. The Omnidata® biophenometers allow up to five different upper-lower threshold accumulations to be stored on a single biophenometer. The Omnidata biophenometer allows the upper cutoff to only be of the horizontal variety.

The starting point for the heat accumulation is the "biofix." The biofix is some easily observed event, such as the first moth catch, which is used to synchronize the model with the field populations.

FIGURE 2

Plot of developmental rate versus temperature. This plot is used to determine the total degree-days to complete a stage and the lower threshold for development.

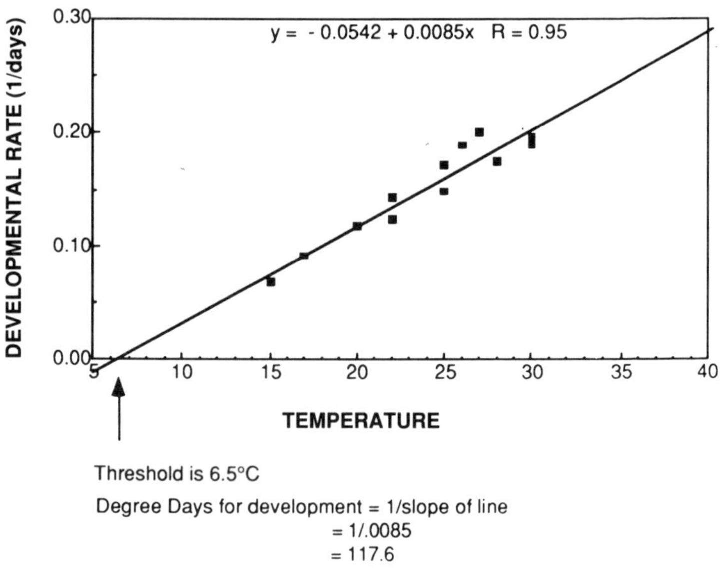

Occasionally, it is a calendar time, such as March 1, before which you generally get no heat unit accumulation. This allows to you to set up your weather equipment later and to save time and effort early in the season when little of interest in the insect life cycle is occurring.

DETERMINING DEGREE-DAYS AND THRESHOLDS

How are the number of degree-days for development of the insects determined? Normally, groups of 30 or more insects are reared in the lab at a constant temperature, and the length of time required for each individual to complete each stage (egg, larva, pupa, and adult) is recorded. This is repeated at four or more temperatures with different groups of insects. The developmental rate (= 1/average time to complete development) is then plotted versus temperature and fit using linear regression (*see Figure* 2). The lower threshold for development is where the line drawn through the points intercepts the temperature axis (where developmental rate = 0), and the degree-days required to complete the stage are equal to 1/slope of the line drawn through the graph (the slope is the average rate of development in terms of proportion development completed per day at a given temperature; therefore, 1/slope determines how long [in degree-days] it takes to complete the stage in question).

MODELING EMERGENCE

Now that we know the upper and lower thresholds and the number of heat units to complete each stage, we need to be able to model how the insects emerge. This information is taken from the same laboratory experiments mentioned above—we know the emergence pattern for our laboratory populations from the way we took our data. At this point, we generally choose some statistical distribution which is adaptable enough to adequately model the laboratory data. Several distributions have been used, such as the Weibull and the Erlang distributions, however, the key point is not which distribution is used, but how well the chosen distribution predicts the field population emergence.

For deciduous fruits, the PETE model developed by Brian Croft and his co-workers uses the Erlang distribution. This distribution was used because it is skewed to the right (i.e., a large portion of the population emerges quickly, but a small part tends to emerge later) and also because by changing a single parameter of the model (k), the emergence curve can take on many different shapes (*see Figure* 3). This adaptability means that by simply changing k, a wide variety of insects can be modeled without changing the underlying distribution upon which it is based.

FIGURE 3

Differing shapes of the emergence curves possible using the Erlang distribution. The mean emergence time is the same, only the shape parameter (k) has been changed.

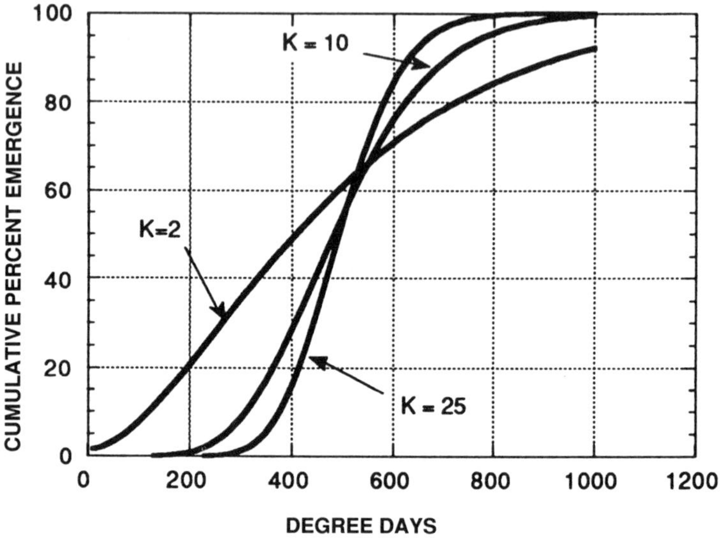

With our laboratory information now modeled, we take the model to the field. Development times per se are rarely measured, instead, we measure the portion of the population in each developmental stage and compare it to the temperatures observed in the field. The process of validation can take several forms, but for IPM, the idea of an "indifference band" validation is frequently used. The indifference bands are a period in time about the predicted event where if the model is off during this period of time, it will make no difference to the pest management program. These bands are set depending on the biology of the pest and the effect of applying the management tactic early or late on its eventual success.

To determine the failure rate of the model, you plot the date an event was predicted to occur and when it actually occurred in the field population. If the point falls outside the indifference band, it is considered a failure; otherwise, it is a success. The percent of the time the model fails is thus an indication of how useful the model is at predicting pest management decisions. Of course, this must be done over a period of several years and in a number of different orchards to ensure success. This sort of plot is shown in Figure 4 for the Western cherry fruit fly information from both Utah and Washington.

FIGURE 4

Indifference band validation for Western cherry fruit fly on cherries in Utah and Washington.

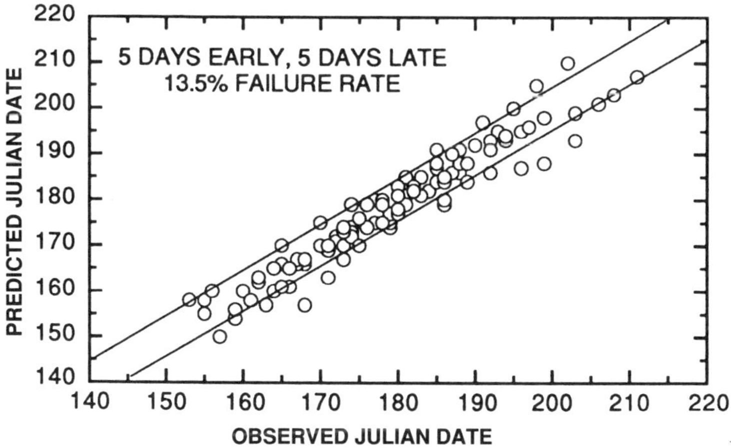

EXAMPLES OF SUCCESSFUL MODELS

There are at least nine insect models applicable to deciduous fruits: codling moth, white apple leafhopper, Western cherry fruit fly, San Jose scale, peach twig borer, Oriental fruit moth, tentiform leafminer, apple maggot, and oblique-banded leafroller. Most of these are available in Washington State or will be soon.

Probably the most successful model is the codling moth model. It is used in most fruit growing regions of the United States to time the first and third cover sprays. The model is used to predict when egg hatch occurs, based on the amount of heat accumulated since the first consistent flight of codling moth (i.e., the biofix). The goal of the cover sprays is to kill the young larvae before they begin to bore into the fruit, and, secondarily, to kill adult moths present in the orchard. The use of the codling moth model allows us to get the maximum effective longevity of the cover sprays.

It is critical that the sprays be applied as close to the time predicted as possible. Putting the spray on too early means the residue has weathered that much before the larvae emerge and must be applied that much sooner. Putting them on late means more larvae have entered the fruit than desired. A mistake of applying the pesticide too late results in damaged fruit at harvest or early fruit drop. Using the older methods of 21 days after first bloom, Beers and Brunner have found the first cover spray went on an average of seven days before egg hatch began (and up to 18 days early), thus wasting a significant portion of the 21-day longevity of the Guthion residue.

Current recommendations in Washington State call for the first cover spray to go on 250 degree-days after the biofix. This corresponds to approximately three percent egg hatch. The second spray goes on based on the 21-day longevity of the Guthion residue, and the third spray goes on at 1260 degree-days or when about seven percent egg hatch of the second generation has occurred. Why apply at seven percent instead of three percent again? Well, if your first two applications were properly timed, then the in-orchard population will be considerably smaller and seven percent egg hatch will still not be that much damage at harvest.

In Utah, growers were complaining of codling moth resistance problems in the early 1980's. The recommended rates of Guthion were increased to about one-and-a-half to two times higher than the rates used in Washington State, but damage did not decrease. Instead, mite problems became much more severe, and predator mites were rarely found following a cover spray. When I arrived in 1985, I surveyed several of the orchards for resistance, but did not find any. A close look at the timing of the sprays revealed that the timings from the codling moth model were very far off from predictions from New York and Washington State (*see Table 1*). A misinterpretation of the model had caused the extension service to recommend the

TABLE 1

Predicted codling moth egg hatch using models developed in New York, Utah and Washington State (from Jones 1987).

% Egg Hatch	Utah Old	New York	Washington	Utah New
1	280 (4.4)*	209	220	225
5	420 (30)	273	270	285
10	490 (47)	304	320	320
25	625 (73)	382	380	400

**Numbers in parentheses indicate the % egg hatch according to new interpretation.*

first sprays be applied at 30 percent egg hatch rather than the desired five to seven percent egg hatch. By changing the timing to be similar to Washington State's, the rates of Guthion recommended were dropped to the correct levels, control was re-established, and mite problems decreased.

Another model which shows several points of interest is the

apple maggot model I worked on for Utah tart cherries. Apple maggot has been a pest in the East for about 100 years and has been the subject of many papers which mention its phenology. If you look at the average time of emergence from 152 different location/years worth of data, you find the mean calendar date of emergence is June 23. However, the average emergence time in different locations/hosts/years can vary from June 2 (Canada) to July 18 (Oregon). Obviously, calendar dates don't provide accurate enough predictions for IPM.

Use of the heat unit concept should clear this up, right? Well, not as much as we'd like to think. If you look at the heat accumulations since the first of March for studies where it is available, first emergence varies from 563 degree-days (New York) to 965 degree-days (Oregon). This is not due to accumulations occurring before the biofix, but is probably a result of differences in growing conditions (particularly soil moisture), orchard aspect, and ground cover between regions, and, possibly, in the stage of development of the overwintering pupae that have entered diapause.

Simply using a model developed in one location without validation is a major mistake. We found that the model could be used to predict the emergence curve (i.e., the rate of emergence) in Utah, Canada, New York, Washington, and Oregon, but failed to predict the Wisconsin emergence curve. The model developed for Utah flies resulted in a 77.5 percent failure rate when using data from New York for validation. However, the apple maggot model developed in Utah appears to predict populations emerging in Washington State, according to some of Jay Brunner's data.

So far, I have basically been discussing the importance of models for timing purposes. However, they are also excellent tools for looking at the interactions between different pests. For example, the second codling moth cover spray occurs very close to apple maggot emergence in Utah. Although apple maggot does not currently attack apples in Utah, we investigated the interaction between the two pests in preparation for when apple maggot moves over to apple. The two insects have different thresholds for development, so the difference in emergence times between the two populations changes with each year and between different geographical locations. By using real-time weather data, we found that in most years and locations, we could adjust the spray schedule for both pests slightly to result in one less cover spray in areas where both would occur. This can be expanded further by looking at more pests to see how the different control measures interact.

In summary, models are tools which help decide when the

optimal time of pest control occurs. They are based on the insects' growth being closely correlated with the ambient temperature and are best used to predict when some difficult-to-observe event has occurred. It should be noted, however, that models are not a replacement for sampling. In fact, when used to predict the optimal time for sampling (e.g., when second generation of tentiform leafminer has begun), it can add to the information received by sampling and further provide information on what the population may look like if little predation occurs or no controls are applied (e.g., if you expect 10% to have emerged by a given time and you've caught nothing in your traps, it will probably be a light year for that pest).

This is journal series article no. 3566 of the Hawaii Institute of Tropical Agriculture and Human Resources, University of Hawaii at Manoa.

8

SAMPLING FOR INTEGRATED PEST MANAGEMENT

VINCENT P. JONES
University of Hawaii
Honolulu, Hawaii

THE TITLE OF THIS CHAPTER, "SAMPLING FOR IPM," implies that sampling is important for some reason. Indeed, often the researcher or extension agent assumes that the importance of sampling is obvious, and they mention it only briefly within a sentence with some catchy phrase like "it helps you determine whether you need to spray." While this is in fact true, it does not explain why you need to collect 20 leaves from 10 trees in various parts of the orchard rather than 10 leaves from 20 trees, etc. In this chapter, I will explain the jargon of sampling, why sampling is necessary, how frequently you need to sample, why sampling plans are not as simple as practitioners would like, and how sampling plans can be made more simple, time-efficient, and accurate. In addition, I will finish the chapter with three examples: spider mites, *Typhlodromus occidentalis*, and western tentiform leafminer.

WHY SAMPLE?

The simplest reason why you need to sample in a pest management system is so that our decisions are based on the best information possible. It is unreasonable to expect a pest management program to work if you are basing it on false premises. For example, if your economic threshold is set at eight mites per leaf and your sampling plan tells you the average number of mites per leaf is five, you should not spray. However, if the sampling plan is poorly designed and you actually have 12 mites per leaf, you have made an error in the management of your crop and allowed economic damage to occur. Conversely, you could actually have five mites per leaf, and your sampling plan could tell you that you have 10 mites per leaf, and this causes you to spray when it is unnecessary. In both cases, the economic threshold and your management plan are not incorrect, your sampling program is. People who work with computers have a acronym which describes this sort of problem—GIGO (garbage in = garbage out).

In pest management, there are many times when sampling should be carried out. These include: 1) when the pest is not feeding directly on the marketed commodity (e.g., spider mites, tentiform leafminer); 2) when the pest has tremendous potential to increase during a short period of time; 3) when you need to have accurate estimates to determine whether the natural enemies will suppress the population of the pest or whether the grower must spray; 4) to evaluate the efficacy of the pest control action (i.e., before-after comparisons of populations); and 5) when an accurate economic injury level for your pest is available. Having accurate estimates of pest populations also allows you to look at the growth rate of the insect pest from generation to generation and decide if control will be required or not. Data should also be saved from year to year to allow you to fine tune your management program by seeing how decisions made previously worked out compared to your current situation.

HOW OFTEN SHOULD I SAMPLE?

The frequency of sampling depends heavily on the biology of the pest and its natural enemies. For example, the western tentiform leafminer has three to four generations per year. The potential for population growth from generation to generation is quite high, but a generation may take a month or more to complete. It does not pay to be sampling western tentiform leafminer more than once per generation (sample near the end of each generation). Samples taken at weekly intervals will show little change and, hopefully, reflect only the acceptable variation in your sampling plan. On the other hand, spider mite populations can complete a generation in seven to 10 days under high temperatures in the middle of summer. Sampling should be carried out in these situations at approximately weekly intervals, so that you are continually aware of what is happening before the economic injury level is surpassed.

Statistical Distributions and Why Samples of a Certain Size Are Important

If you take a sample of 10 leaves and find that there are 30 mites present, the average number of mites per leaf is 30/10 or three mites per leaf. The mean gives you an important piece of information about the population, but you can have several markedly different distributions of the mites among the 10 leaves. By tallying the number of leaves which have 0, 1, 2, 3,... mites present, you construct a frequency distribution. Figure 1 shows three common types of frequency distributions, each of which has a mean of three mites per leaf, but each of which has a markedly different variance about the mean population level. In the uniform distribution pictured, taking even one or two leaves gives a close estimate of the mean population level, and the low variance emphasizes this point. Unfortunately, this sort of distribution almost never occurs with insects or mites.

FIGURE 1

Frequency distributions for three hypothetical mite populations, all with a mean of three mites per leaf, but with different variances.

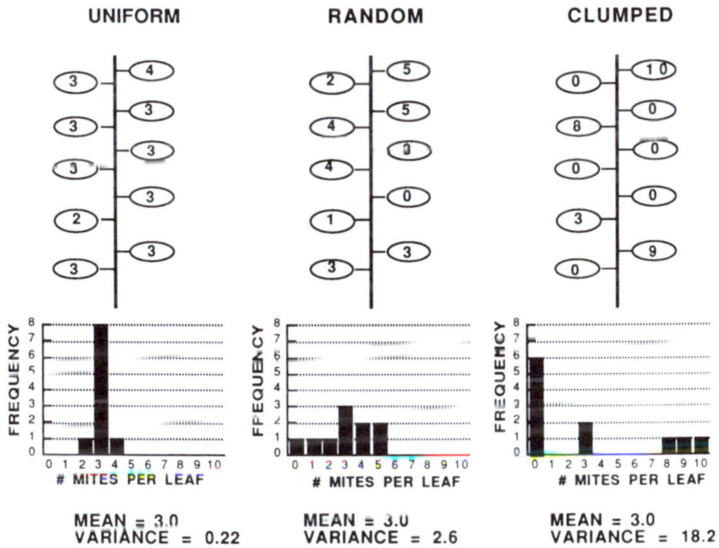

The other extreme is the clumped distribution. In this situation, most of the leaves have no or only a few individuals present, while a few leaves have most of the individuals in the sample. If you took a random sample of three leaves from our example distribution and calculated the average, the estimate could vary from nine mites to zero mites per leaf. This sort of error in the estimation of the mean population causes errors in the management plan because your sampling plan is inadequate. Unfortunately, this sort of distribution is characteristic of insect populations.

The ratio between the variance and the mean population level is important in determining the shape of the distribution and for deciding how many samples must be taken to give an estimate of the population level which has the desired accuracy. However, the relationship between the mean and variance changes as the mean population level changes. Fortunately, it changes in a manner which can be predicted and tends to be consistent for mites with similar demographic characteristics across a wide variety of hosts. If you have a large enough collection of samples, you can plot the loga-

FIGURE 2

The relationship between the mean of a sample and its variance can be predicted by linear regression. Data shown are for two-spotted spider mites on seven different crops, including apple, peach, and tart cherry.

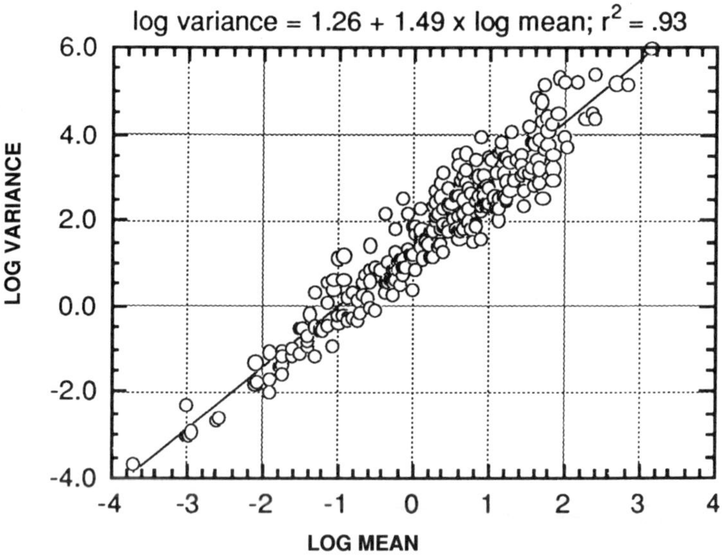

rithm of the mean against the logarithm of the variance for all the samples. When you do this, you end up with a straight line that allows you to predict what the variance is from just an estimate of the mean population level. Figure 2 shows this type of plot for two-spotted spider mite on seven different host plants, including apple, tart cherry, peach, carnation, rose, and tuberose.

Knowing the relationship between the mean and variance also allows you to model what the frequency distribution is for any given population level. Using this relationship, we can predict how many leaves need to be sampled to give the desired precision or accuracy of our estimates. Perhaps of even more importance, though, is it allows us to construct presence-absence sampling programs. For example, if we solve equations to determine what proportion of leaves have mites on them at a range of densities likely to be encountered in the orchard, you can estimate the mean population level from just tallying the number of the leaves that have one or more mites present. To show how well this works, you can plot the proportion of leaves infested and the mean population level obtained from counting all the mites present (*see Figure 3*).

The problem with this type of sampling plan is that estimates of the mean population level when the proportion infested increases

FIGURE 3

The proportion of leaves can be used to predict the mean population level using information known about the distribution and the mean/variance relationship shown in Figure 2.

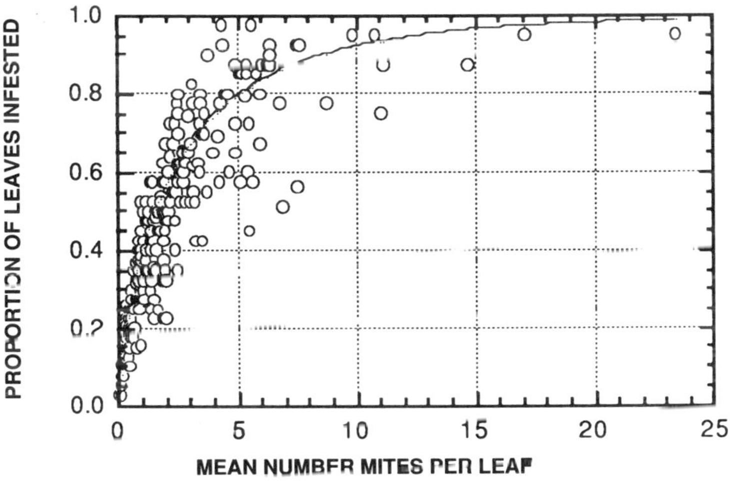

above about. 0.80 to 0.85 are often unstable. However, the presence-absence concept can be extended by looking for the presence-absence of some easily recognized number of insects or mites. For example, using the same relationship between mean and variance which allowed us to predict what the frequency distribution looked like for presence-absence sampling, we can predict the relationship between the mean and the proportion of leaves which have, say, two or more mites present. This means that the highest mean which can be predicted accurately is about 10.6 versus 6.7 mites per leaf with the normal presence-absence sample *(see Figure 4)*. If this level is not high enough, the proportion of leaves with greater than 2,3, etc., leaves can be used. However, at some point, the time saved with the binomial sampling decreases, because you need to make sure you have more than your tally threshold of mites present.

FIGURE 4

The relationship between the mean number of mites per leaf and the proportion of leaves infested with more than one mite for two-spotted spider mite and the McDaniel spider mite.

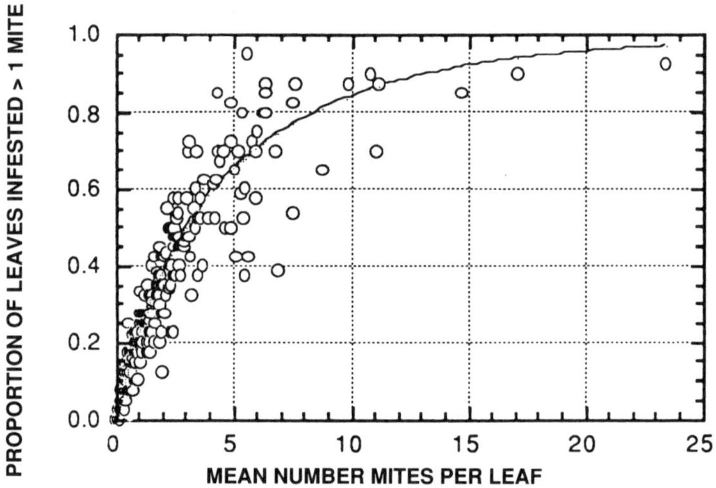

Sequential Sampling

Knowing the distribution also allows us to set up a sequential sample. Formally, a sequential sample is when the total number of leaves in a sample is variable and depends on whether or not the results so far give a definite answer as to whether the economic threshold is surpassed or not. This can be done by either directly counting the number of insects or mites or combined with binomial sampling. For monitoring purposes, it is often easier to explain and use if we ask you to sample a fixed number of leaves from a fixed number of trees, which will provide the desired accuracy when the mean population level is near the economic threshold.

EXAMPLES

Two-spotted and McDaniel Spider Mites

The preceding figures show the binomial sampling information for two-spotted spider mite and McDaniel mite on seven different host plants; this same information and an example are given in Table 1. A total of 25 leaves per tree is required for each of 10 trees per block. Tally the number of leaves which have more than one mite present. Look up the number of leaves infested in column one and follow it across to the third column where the mean population level on that tree is found.

Sample the 10 trees, recording the mean population from each separately, and then take the average of the 10 trees sampled per the block. What does "per block" mean? A block is defined as the smallest unit of the orchard you would consider treating separately. I realize this is rather vague, but until work on the geostatistics of insect and mite distribution in orchard crops (i.e., how are the population levels on trees which are different distances apart related to each other?) is carried out, that is about as precise as we can be. In deciding on your sampling area, you should sample the edges of the blocks separately, since they often have a higher number of pests present than on the internal portion of the orchard.

Typhlodromus Occidentalis

The sampling plan for *T. occidentalis* has been validated to work on apple, peach, and tart cherry. However, I expect that its usefulness is much wider, because values given by researchers in California on almonds are virtually identical to the values I found on the above crops. A presence-absence sample can be used according to Table 2 with 25 leaves per tree. A total of 10 trees should be used for this sample. The sample can be performed at the same time as the sample for two-spotted spider mite.

Western Tentiform Leafminer

The sampling plan for western tentiform leafminer is one where the second generation should be sampled. The present recommendations for Washington State indicate that you treat if more than two mines per leaf are present and there is no parasite activity. The binomial sampling table (*see Table 3*) is based on looking for the presence or absence of more than one mine per leaf on 20 leaves per tree. A total of 10 trees is required to provide accurate orchard-wide population estimates.

CONCLUSIONS

Sampling is not intentionally made harder than necessary. The sample sizes recommended are based on statistics which enable us to accurately predict the population sizes. Inaccurate estimates of population sizes provide poor data, which then leads to mistaken management practices.

TABLE 1

Binomial sampling table for the two-spotted spider mite and the McDaniel mite.

No. leaves infested > 1 mite	Proportion leaves infested > 1 mite	Mean no. mites per leaf
1	.04	0.21
2	.08	0.35
3	.12	0.51
4	.16	0.69
5	.20	0.87
6	.24	1.08
7	.28	1.30
8	.32	1.55
9	.36	1.82
10	.40	2.12
11	.44	2.44
12	.48	2.81
13	.52	3.21
14	.56	3.67
15	.60	4.18
16	.64	4.77
17	.68	5.45
18	.72	6.26
19	.76	7.24
20	.80	8.51
21	.84	10.12

To use the table, collect a sample of 25 leaves and count the number of leaves that have more than one mite present. A total of 10 trees must be counted.

EXAMPLE:

The first tree has 15 leaves with more than one mite present, the second with 10, the third with 7, and the fourth through tenth trees with 18, 6, 5, 17, 10, 11, and 12, respectively. Use Table 3 to estimate the mean number of mites per leaf for each tree and then average across the 10 trees.

Tree	No. leaves infested	Mean population level
1	15	4.18
2	10	2.12
3	7	1.30
4	18	6.26
5	6	1.08
6	5	0.87
7	17	5.45
8	10	2.12
9	11	2.44
10	12	2.81
	Average =	2.86

JARGON EXPLAINED

Definitions:

1. Economic injury level: the lowest number of insects which will cause economic damage, expressed in the number of insects per leaf or per plant part, but actually a level of damage.

2. Economic threshold: the population level at which control measures need to be applied to prevent an increasing pest population from reaching the economic injury level.

(a) The economic threshold is lower than the economic injury level to permit sufficient time for initiation of control measures and for these measures to take effect before the population reaches the economic injury level.

3. Sampling unit: a leaf, a trap, a fruit. Formally, the smallest portion of the environment that is collected for estimation of a population level.

4. Sample: a collection of sampling units that are used to estimate the desired population level. The number of sampling units in a sample is called the sample size.

TABLE 2

Binomial sampling table for *Typhlodromus occidentalis*.

No. leaves infested > 1 mite	Proportion leaves infested > 1 mite	Mean no. mites per leaf
1	.04	0.04
2	.08	0.10
3	.12	0.16
4	.16	0.22
5	.20	0.29
6	.24	0.37
7	.28	0.46
8	.32	0.55
9	.36	0.65
10	.40	0.75
11	.44	0.87
12	.48	1.00
13	.52	1.14
14	.56	1.30
15	.60	1.47
16	.64	1.67
17	.68	1.89
18	.72	2.15
19	.76	2.45
20	.80	2.84
21	.84	3.32

To use the table, collect a sample of 25 leaves and count the number of leaves that have more than one mite present. A total of 10 trees must be counted.

collected in a sample divided by the number of sampling units. The mean is the most commonly used population parameter in decision-making for pest management.

6. Variance: the variation in the number of individuals from leaf to leaf around the mean population level. For example, at an average of five mites per leaf, some leaves may have no mites present, whereas other leaves may have 20 or more mites present. The variance is also sometimes referred to as the "dispersion" of the population (note "dispersion" therefore is different than "dispersal," which refers to movement in and out of the crop).

7. Precision: indicates the repeatability of your sample. It is generally designated as a ratio which reflects the mean and its variation.

EXAMPLE:
The first tree has 7 leaves with more than 1 mite present, the second with 3, the third with 7, and the fourth through tenth tree with 9, 6, 5, 12, 10, 11, and 14, respectively. Use Table 3 to estimate the mean number of mites per leaf for each tree and then average across the ten trees.

Tree	No. leaves infested	Mean population level
1	15	1.47
2	10	0.75
3	7	0.46
4	18	2.15
5	6	0.37
6	5	0.29
7	17	1.89
8	10	0.75
9	11	0.87
10	12	1.00
	Average =	1.00

8. Frequency distribution: if you take a sample of 10 leaves and tally how many of the leaves have no mites present, how many have one mite present, etc., you create a frequency distribution. The shape of frequency distribution gives you an intuitive feel for how many sampling units needs to be collected, so that you are able to accurately assess the mean population level. The types of distribution are:

(a) Uniform: each leaf has approximately the same number of insects present. Basically, only a few leaves would have to be sampled to give you an accurate estimate of the population level. Unfortunately, insects almost never have this distribution.

(b) Random: each leaf has an equal chance of having an insect and the presence of an insect has no influence as to whether another insect is present. Harder to sample than a uniform distribution, but also very rare for insects or mites.

(c) Clumped: numerous leaves have no or only a few individuals present, and only a few leaves have large numbers of insects present. Most common type of distribution for insects and mites, and the most difficult to sample. A large sample size is necessary, so that the few leaves with a large number of insects are included in the estimate.

TABLE 3

Binomial sampling table for the western tentiform leafminer.

No. leaves infested > one mine	Proportion leaves infested > than 1 mine	Mean no. mines per leaf
1	.05	0.33
2	.10	0.51
3	.15	0.67
4	.20	0.81
5	.25	0.96
6	.30	1.10
7	.35	1.25
8	.40	1.40
9	.45	1.57
10	.50	1.73
11	.55	1.91
12	.60	2.11
13	.65	2.33
14	.70	2.57
15	.75	2.86
16	.80	3.21
17	.85	3.66
18	.90	4.28

To use the table, collect a sample of 20 leaves and count the number of leaves that have more than one mine present. A total of 10 trees must be counted for population predictions accurate within one mine per leaf.

EXAMPLE:
The first tree has 5 leaves with more than 1 mine present, the second with 10, the third with 7, and the fourth through tenth tree with 8, 6, 5, 7, 10, 11, and 12, respectively. Use Table 3 to estimate the mean number of mines per leaf for each tree and then average across the 10 trees.

Tree	No. Leaves infested	Mean population level
1	5	0.96
2	10	1.73
3	7	1.25
4	8	1.40
5	6	1.10
6	5	0.96
7	7	1.25
8	10	1.73
9	11	1.91
10	12	2.11
	Average =	1.44

This is journal series article no. 3565 of the Hawaii Institute of Tropical Agriculture and Human Resources, University of Hawaii at Manoa.

9

BIOLOGICAL CONTROL: CONCEPTS AND FUTURE PROMISE

THOMAS R. UNRUH
United States Department
of Agriculture
Agricultural Research Station
Yakima, Washington

BIOLOGICAL CONTROL IS THE REGULATION BELOW damaging levels of pestiferous living organisms through the predatory, parasitic, or competitive activity of beneficial living organisms. This definition excludes many ecologically rational methods of control, including those based on the byproducts of organisms, such as their hormones, pheromones, toxic proteins, or products such as Ryania and avermectin with which field advisors are familiar. I define biological control so narrowly to highlight the management of a multi-species, multi-population interaction. A major goal of tree fruit entomologists is to infuse more biological control into our pest management systems.

Three major types of biological control exist, and they reflect the amount of energy input to create a balance between pest and beneficial populations. These are classical, inoculative, and inundative.

Classical biological control is the practice of seeking important natural enemies of a pest in its native range and introducing these into a new area where pest abundance is unregulated. Inoculative biological control is the periodic or seasonal releases of beneficials to re-establish a balance that has been disrupted. Inundative biological control is the massive production and release of beneficials to quickly produce pest control. In most of what follows, I will address classical biological control.

The premise underlying biological control is that pest species have enemies that, given the correct situation, are capable of regulating their hosts below damaging levels. Examples supporting this assertion abound. They come from naturally occurring biological control and from dramatically successful introductions of natural enemies against exotic pests. A hypothetical example of population

FIGURE 1

A hypothetical example of pest population densities before and after the introduction of a regulating natural enemy. The pattern depicted is typical of many successful biological control introductions. Also illustrated is how the probability of attaining successful biological control—suppression of the pest below the economic injury level—can be a function of the pest's biology. Specifically, direct pests, those that attack the utilized plant parts, are pests at much lower densities than are indirect pests, those that attack plant parts that are not harvested.

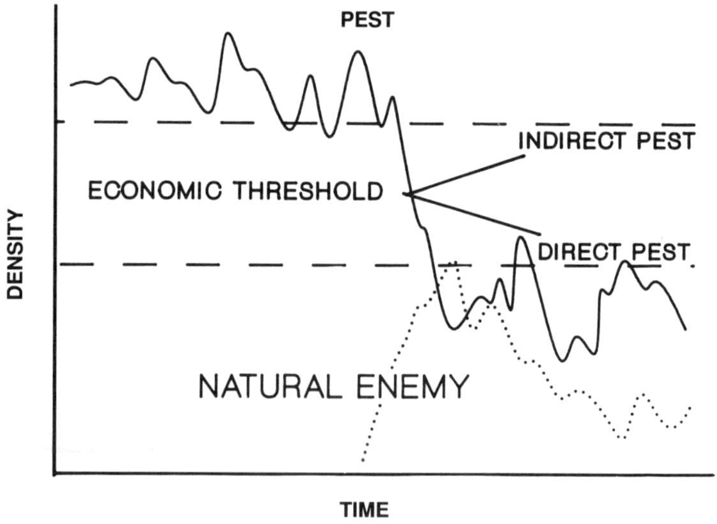

densities of a pest before and after the introduction of a natural enemy is presented in Figure 1. A third form of evidence comes from long-term ecological studies of pests and their natural enemies. But, ecological studies can be misleading, as the following example illustrates.

The winter moth, a polyphagous pest of oak, apple, and myriad other deciduous trees, is probably the best studied of all classical biological control efforts. Winter moth colonized and became a pest in Nova Scotia and on Vancouver Island of British Columbia. Dramatically successful control followed the introduction of two of the moth's natural enemies, the parasitic fly *Cyzenis albicans* and the parasitic wasp *Agrypon flaveolatum*. Surprisingly, in England, part of the moth's native range, two decades of basic ecological studies indicated that these parasites were ineffective in regulating the moth. This discrepancy between basic ecological research and introductions points out the difficulty of actually measuring what regulates natural populations. Biological control introductions are, in effect, grand ecological experiments. When introduction of a beneficial agent controls a wayward insect, the regulating factor is demonstrated. Unfortunately, biological control is seldom a predictive science. Another example of winter moth control may soon be recorded. This species is approaching outbreak levels in northwestern Washington State. We surmise that the moth recently colonized the mainland from Vancouver Island.

Natural biological control often becomes evident when the action of natural enemies is disrupted by pesticides or by more or less natural disturbances such as road dust, the interfering activity of ants, or extremely cold winters. Direct physical disruption and exclusion of natural enemies have proven their role in regulating pests. Cages or screens are used to keep pests in and their natural enemies out. Invariably, tremendous population growth of the protected pest occurs, compared to control cages which allow access by natural enemies. In one classic study, all mite and insect predators and parasites were manually removed from an avocado tree for 84 consecutive days. Pest mite populations increased slightly, then declined in surrounding orchard trees, but the mite population on the predator-free tree increased to and stayed at very high, damaging levels until the end of the study (*see Figure 2*).

Other evidence of natural control in tree fruits comes from the disruption of beneficials by pesticides. A complex of mealybugs exploded in pears in California after trees were treated with DDT in the early 1960's. The cause of this upset was the destruction of lacewings that normally kept the mealybugs in check. The problem was solved by periodic inoculation of orchards with lacewing eggs, since the immature stages of lacewing were tolerant to DDT.

FIGURE 2

Natural enemy removal experiment. All mite and insect predators and parasites were manually removed from an avocado tree for 84 consecutive days. Shown is avocado Brown mite density on this tree (predators removed) and in the unmanipulated trees in the surrounding orchard (predators present).

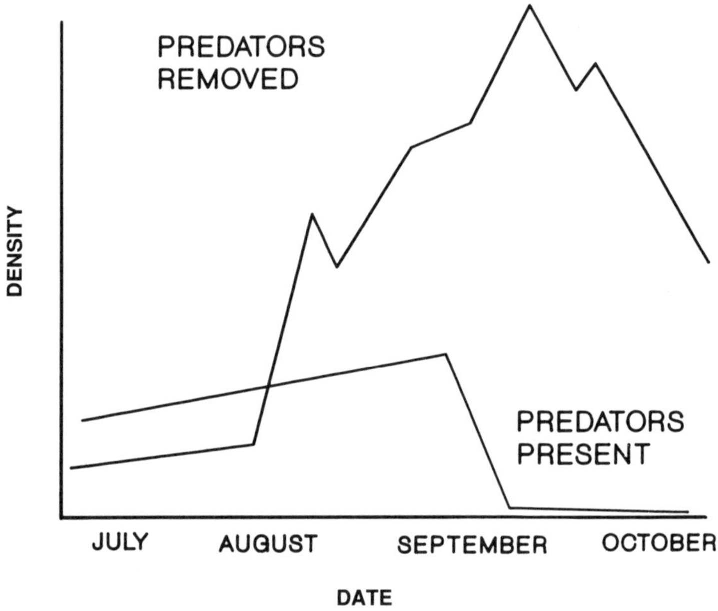

Insects that become pests because their natural control agents are killed by pesticides are described as pesticide upset pests. Several secondary pests of tree fruits are upset pests. Tetranychid mites are upset by pyrethroid use; Western tentiform leafminer are upset by broad spectrum pesticides, as is white apple leafhopper. Crop damage by pests such as San Jose scale and woolly apple aphid may be due, in part, to overzealous and poorly timed pesticide use. A pest of primary importance, pear psylla, also is an upset pest, or, at the very least, its severity is exacerbated by the broad spectrum pesticides commonly employed for codling moth control. I will return to upset pests and codling moth control later.

Other dramatic evidence of the impact natural enemies may have in regulating host abundance comes from accidental introduction of detrimental organisms. Probably the most notable example in North America is the decimation of American elm by Dutch elm disease. At this point, one should consider that "pest" and "beneficial" are quintessential anthropomorphic terms. Biological traits we use to classify an insect as a pest of fruit trees may be desirable in a biological agent for weed control.

WHAT MAKES A GOOD NATURAL ENEMY?

Attributes used to measure the value of biological control agents include: adaptation to, and preferably specialization on, the target pest; inability to interfere with the action of other established beneficial insects; a well-developed ability to find its host and utilize it efficiently; and climatic tolerances suited to its new home. Each of these invoke common sense and are supported by observed patterns of success.

Climatic Matching

Climatic matching of the natural enemy to its new home was key to the successful biological control of the walnut aphid in California. The first introduction against this Eurasian pest was the parasite *Trioxys pallidus* from France. It gave substantial control in the cool southern and coastal areas of California but was ineffective in the more extreme climate of the Central Valley. Several years later, a race of *Trioxys pallidus* was imported from Iran, which is climatically very similar to central California. The Persian race gave excellent control of walnut aphid in the Central Valley.

Specialization on the target pest by the natural enemy, in retrospect, appears critical to success in many systems. The woolly apple aphid is well controlled in most countries by its species-specific parasite *Aphelinus mali*; this parasite makes its living exclusively on the woolly apple aphid. Sometimes, specialized races or closely related beneficial species exist which differ substantially only in their preference for a particular host. For example, the strains of *Trioxys pallidus* imported for walnut aphid were ineffective in controlling the filbert aphid when they were released against it in Oregon. However, discovery and subsequent introduction of a filbert aphid-attacking race of *Trioxys pallidus* resulted in excellent biological control.

The number of natural enemies acting on a specific pest is highly variable, and one concern of biological control practitioners is the possibility of interference between natural enemies resulting in control that is worse than either natural enemy alone. Fortunately, there are many examples of two or more beneficial species working

FIGURE 3a

The four panels depict parasitism rates of pear psylla by nymphal instar at a soft pesticide orchard (Tieton) and an unsprayed research orchard (YARL) for two consecutive years. Psylla-infested foliage was collected in the field, and nymphs were removed by instar and reared on greenhouse grown plants until adults parasites or psylla emerged. High parasitism rates in the spring of 1988 at YARL and 1989 at Tieton indicate *Trechnites* can synchronize to and significantly impact spring psylla populations. Low spring rates in the opposite years at each site suggest that an even more tightly synchronized biology may be beneficial in a parasite of psylla.

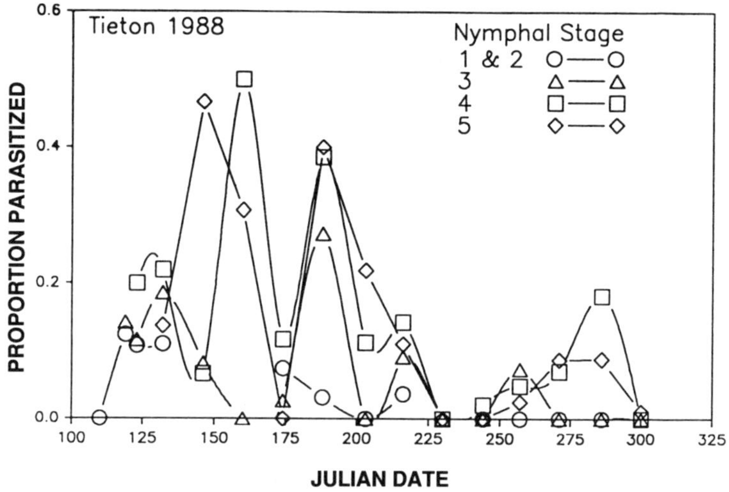

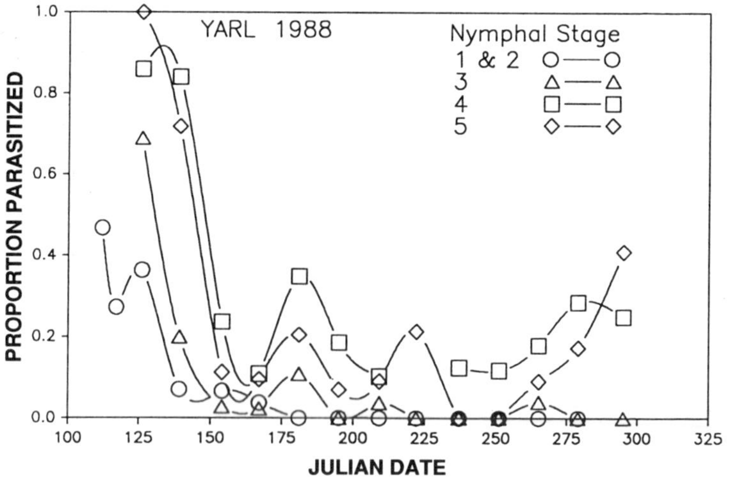

FIGURE 3b

The four panels depict parasitism rates of pear psylla by nymphal instar at a soft pesticide orchard (Tieton) and an unsprayed research orchard (YARL) for two consecutive years. Psylla-infested foliage was collected in the field, and nymphs were removed by instar and reared on greenhouse grown plants until adults parasites or psylla emerged. High parasitism rates in the spring of 1988 at YARL and 1989 at Tieton indicate *Trechnites* can synchronize to and significantly impact spring psylla populations. Low spring rates in the opposite years at each site suggest that an even more tightly synchronized biology may be beneficial in a parasite of psylla.

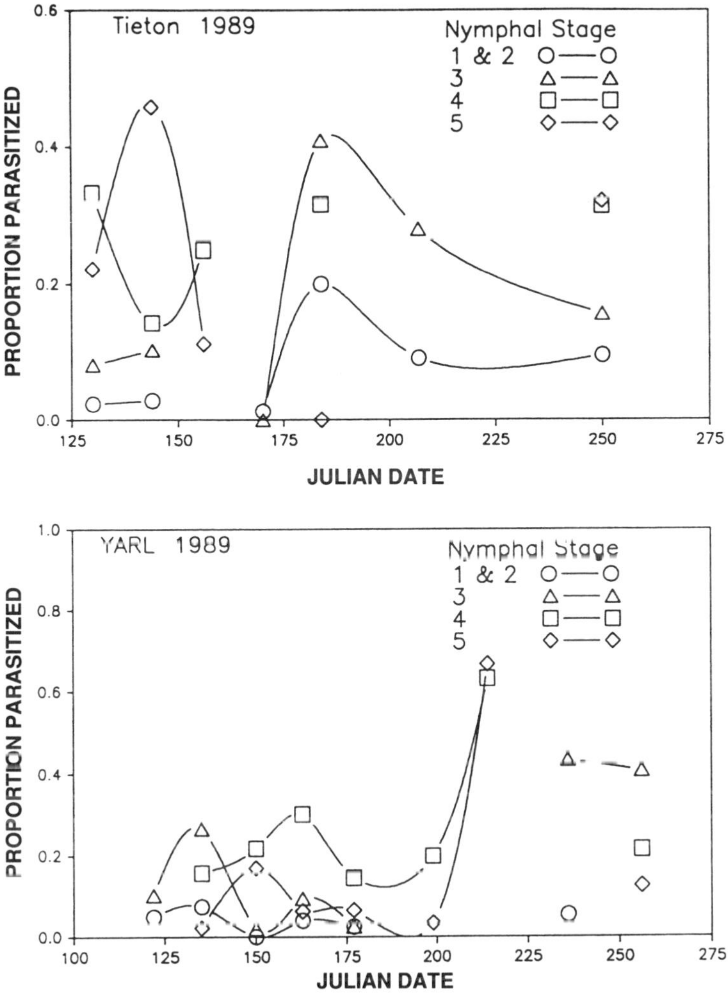

in concert to regulate a pest, while instances of interference are rare. The best example of the need for multiple species introductions is the successful biological control of the olive scale. In the 1950's, several races of the parasite *Aphytis maculicornis* were introduced from various climatically different sources in Eurasia. A strain from Iran established and provided partial control of the scale. While it killed most of the scale in the cool springs of the Central Valley, it could not withstand the hot dry summers. In the late 1950's, the parasite *Coccophagoides utilis* was imported from Pakistan, and it added significant parasitism in the summer generation, bringing the scale under complete control.

The walnut aphid and olive scale examples also illustrate the flip side of climatic matching. That is, if the natural enemy is not adequately adapted to its climate, low winter temperatures or hot summers may prevent control of the target pest. Thus, ineffective control of several pests, including some in tree fruits, may be caused by maladapted natural enemies and may be correctable by introduction of better adapted ones.

Host Synchrony

An equally important aspect of climatic matching is host synchrony between the population cycles of predator and prey. My work with pear psylla has focused on one of its natural enemies, the parasite *Trechnites psyllae*. Synchrony is critical in this interaction. The beneficial wasp stings and lays its egg in the second through fifth nymphal stages of psylla. The parasite larvae develops inside the nymph for about two weeks and mummifies and kills the nymph. Only one wasp emerges from a single psylla. Synchrony is critical in the overwintering period. Psylla overwinter as adults more or less outside of pear orchards and return in early spring to lay eggs. In contrast, the parasites developing in psylla nymphs in fall induce their host to crawl off the leaves into bark crevices where they mummify. Thus, the parasite overwinters in the orchard as diapausing larvae inside this hidden psyllid mummy.

The exact commencement time in spring for egg laying by psylla varies considerably from one year to the next. *Trechnites* emergence in spring must be synchronized to this moving target to control psylla, at least for the first summer generation. Significantly, my studies in unsprayed pear orchards suggest that *Trechnites* can kill a significant proportion of this spring brood of psylla during a time when predators are notably absent (*see Figure 3*). However, at some locations in some years, *Trechnites* may miss the mark. A parasite with greater synchrony with pear psylla under our conditions may be valuable.

Functional Response

Efficiency in host finding and utilization represent a complex of traits which have been the focus of much research. The behavior most often studied is the functional response. Functional response is a measure of behavioral changes made by a predator in response to different host densities. The critical question raised in studying functional response is: does the pattern and rate of host destruction lead to regulation of the pest?

As in systems theory, negative feedback between the predator and host populations is considered essential for stable population interactions. The alternative, instability, means population cycles will be extreme and, eventually, one or both species will go to extinction. In predator-prey modeling studies, and even in laboratory microenvironments, the predator or beneficial population almost invariably is the one which dies out. However, several kinds of

FIGURE 4

Functional response is the change in foraging behavior of a natural enemy in response to changes in density of its host. The Type II response is characterized by an ever decreasing proportion of prey killed with increasing prey density. The Type III response shows an increase in proportion of prey killed over the lower ranges of prey densities, but it reverts to a Type II decrease at the higher prey densities.

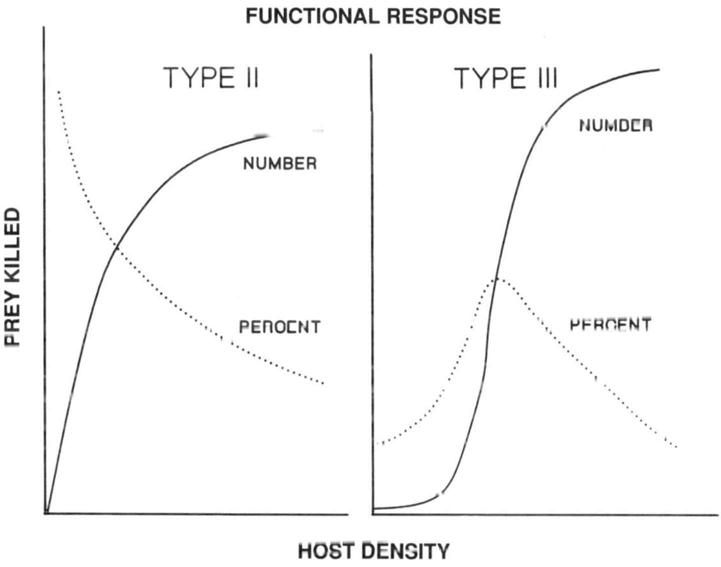

predator behavior can provide negative feedback and stabilize predator-prey interactions. Required is a behavior that causes the proportion of prey killed to increase with increasing prey density.

Two types of functional response are commonly observed in invertebrate predators. The Type II response *(see Figure 4)* shows a negatively accelerating increase in the number of prey killed as their density increases. When transformed to percentage of prey killed, Type II shows an ever decreasing percentage with increasing prey density. The Type II functional response is not thought to stabilize predator-prey model systems, although it is the most commonly observed response in laboratory studies of predatory and parasitic insects. The other commonly observed response is the Type III *(see Figure 4)*. Here, the number of host killed with increasing host densities shows a positive acceleration through the lower host densities, after which it switches over to a negative acceleration for the remaining higher densities. This response shows an increasing percentage killed over the lower range of prey densities. This is a negative feedback and can be stabilizing under some conditions.

We know that behaviors or physiological changes can give rise to a Type III functional response. Host switching is one such behavior. Imagine a predator population feeding almost exclusively on a common host until another host increases to a density sufficient to make it more profitable for the predator to feed on it. Such behavioral switches are well documented in generalist predators, including some insects. A second behavior that yields a Type III is a simple giving up time rule. Picture a pest distributed into discrete patches, say leaves of a tree. The predator arrives on a leaf and begins to search. If it fails to find a host after a fixed time, it leaves the leaf. On the other hand, when the predator encounters a host, it tends to stay and search the leaf more thoroughly (longer giving up time after host encounter). Thus, when many hosts are on a leaf, the predator is likely to encounter one and then its behavior will change to an intensive mode, which results in most, if not all, of the prey being discovered.

Insect predators and parasites are usually too small or too unwieldy to study their behavior in a natural setting. Thus, we simplify nature and watch insects behave in artificial laboratory settings. Unfortunately, these simplified worlds can be misleading. For example, if we test a predator at host densities above the range where killing rates increase, we will mistakenly assume a Type II response. Also possible is that laboratory artifacts such as cages may prevent other behaviors that produce a Type III response, such as dispersal or giving up behavior after a failed first look for prey. In fact, trying to predict the value of a potential natural enemy from their laboratory functional response is misguided, because we will never quite be sure how these are expressed in the field. However, functional response and related measurements can allow us to frame suitable questions for testing in the field.

The concept of negative feedback has been important in biological control theory for many years. In the past, theorists were divided into two camps. There were those believing that natural enemies controlled the abundance of most insects and negative feedback, or density dependence, was required. The second camp believed that insect abundance was regulated by natural factors, usually weather, that acted independently of density. In reality, both types of mortality occur, but negative feedback may come from interactions not embodied by the functional response curve. The current focus is on the spatial dimension of population interactions.

Imagine a predator so efficient that whenever it lands on a patch of habitat occupied by prey that it consumes them all. No increase in proportion killed with density will be evident in occupied patches. Also, allow that the prey occupy some patches, or refugia, that the predator can never find. Finally, let the prey be slightly more adept at reproducing and colonizing new patches than the predator is at finding them. This interaction can result in a balance of colonization of new patches by the prey and prey extinction brought on by the predator. Theorists contend that such a process of coupled interactions will yield a stable population of prey and predators. The density level of each is a function of birth rate, dispersal rate, search efficiency, and size of refugia.

BIOLOGICAL CONTROL OPPORTUNITIES

Now, to return to biological control opportunities in tree fruits in Washington State. Dr. Hoyt has ably shown the feeding relationships of mites on apples and how pyrethroid insecticides can induce upset of mite pests. One pest that cannot be accused of being an upset pest is the codling moth. This persistent fruit pest is the trigger for most of our pest management problems in pear and apple, and elsewhere, in walnuts. As a direct pest, one which directly damages the fruit, the economic injury level of codling moth is very low *(recall Figure 1)*. As an exotic species, probably originating in Soviet Central Asia, and with only one of its Old World parasites established here, there is little measurable impact of natural enemies. Thus, for the last 60 years or more, we have controlled codling moth with chemical insecticides to the detriment of biological control of other pests. What are our options with codling moth and what do they portend for pest management in our orchards?

At laboratories in the United States, Canada, Europe, and several countries in the Southern Hemisphere, researchers have demonstrated that synthetic pheromones can be used to disrupt mating and thereby control codling moth. Registration of this synthetic pheromone for orchard use may come none too soon, given recent reports of resistance to azinphosmethyl in various codling moth populations. On a parallel track in British Columbia, planning continues for a large sterile insect program to virtually eradicate the codling moth in the main apple regions of that province. It is too soon to tell if the mating disruption approach may derail those plans. Lastly, soft pesticide programs such as those based on insect growth regulators are being developed and should provide other, nearly species-specific, nondisruptive methods to control key pests and sporadic pest outbreaks. It is my hope that mating disruption becomes fully implemented and is a resounding success for the sake of our growers, workers, and our common environment.

Also, I hope researchers and the granting bodies that support them will switch their attention to improving biological control of the other pests of apple and pear. There are plenty of opportunities here. For starters, pear psylla, which is also from Eurasia, has seen little foreign exploration for exotic enemies outside the easily travelled areas of western Europe. Studies published in the Soviet Union suggest that parasites related to the *Trechnites* we have here substantially control this pest in eastern Europe and central Asia. On a recent visit to these regions, I found it very difficult to find pear psylla even in commercial orchards where climates are similar to our own and where psylla is known to occur. We need to discover what is regulating this pest in these regions, and if it is biological agents, we should import them.

Similarly, additional natural enemies for San Jose scale, woolly apple aphid, Pandemis leafrollers, and selected other minor pests should be sought. Other pests such as the white apple leafhopper and the tentiform leafminer will, I predict, become nonproblems under a mating disruption management program. Our work on monitoring secondary pests in experimental mating disruption and organic orchards supports this contention for the leafminer *(see Figure 5)*.

Finally, the codling moth itself should be the target of beneficial introductions. Many species are known to attack it in the Old World, and in some areas of central Asia, they keep fruit damage around 30 percent in commercial orchards in areas where codling moth completes three generations. Our growing area is climatically very similar, but here, an unsprayed orchard shows virtually 100 percent fruit

FIGURE 5

Examples of the effect of pesticides on the suppression of western tentiform leafminer. Shown are the density of the leafminer in paired blocks at two areas of the Yakima, Washington, growing region. A) A commercial block which uses standard Guthion treatments to control codling moth compared to a nearby orchard that has been under typical organic management (oil, *Bacillus thuringiensis*, Rotenone) for at least 12 years. B) Paired blocks in another orchard; one used conventional pest management, and in the other, codling moth control was by pheromone mating disruption. Parasitism rates were not taken, but I surmise the late season increases in WTLM at both commercial blocks is caused by destruction of parasites and release of the pest from pesticide suppression near harvest.

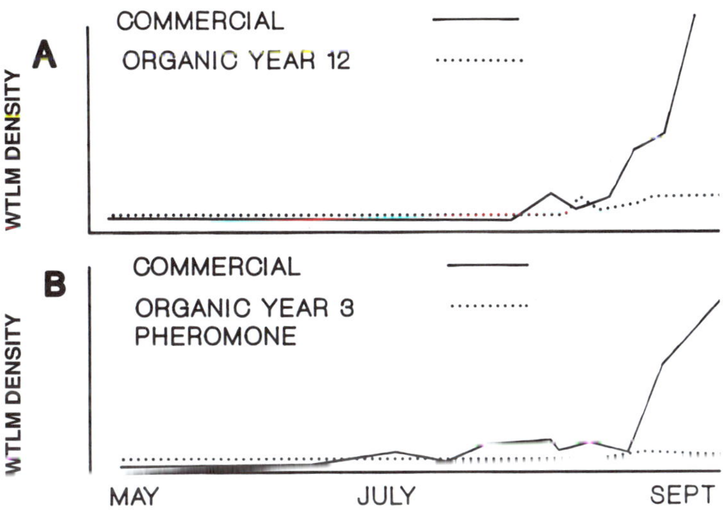

damage, with most fruit having multiple larvae. If nothing else, the establishment of beneficials of codling moth in the extra-orchard setting of crab apple and suburban apple trees, could decrease by half our problems of codling moth entering orchards from these sources.

New Pests

Perhaps the biggest concern for apple and pear production in the next decade is the invasion of our area by new pests. The apple maggot and the plum curculio, both endemic to the eastern United States, are serious possibilities. Although soft management programs have been worked out for the apple maggot, they are expensive and labor-intensive to initiate. The plum curculio is more serious still. It is usually controlled in the East with one to two applications of broad spectrum insecticides. A monitoring program for plum curculio should be maintained by Washington State.

Washington has recently been colonized by two exotic pests of apple—the apple ermine moth and the winter moth. I have been actively involved in a classical biological control program against the apple ermine moth and can report establishment of one of its most important parasites from Eurasia. This pest was detected in outbreak proportions on suburban trees in the Bellingham area in 1985. Since, it has spread down to the Columbia River west of the Cascade Range and has breached the Cascades through two major passes. Populations near Bellingham have stabilized or declined owing to the action of generalist predators such as birds and spiders. This insect is not a threat to chemically managed orchards, but it could be a persistent pest in orchards where codling moth is controlled with mating disruption.

Control of the winter moth by parasites that now control it on Vancouver Island and elsewhere should be possible for our mainland populations. Washington State's Department of Agriculture is involved with Canadian researchers to import and establish these parasites.

Biological control has an expanding future for fruit pest management in the Pacific Northwest. Although Washington fruit growers have accepted and used integrated pest management tactics for more than a generation, a new emphasis on putting more biological control into IPM is going to keep pest control advisors busy. Decisions to spray for a pest will need to be weighed ever more carefully against the potential effects on predators and parasites. These decisions require increased knowledge of the beneficial complex. Pest/natural enemy ratios need to be used in decision-making for pest control; these sorts of decision strategies are available for mites and leafminer. We need to do the research that would allow these decisions for most, if not all, pests of tree fruits.

10

MATING DISRUPTION AS A CONTROL FOR FRUIT PESTS

JAY F. BRUNNER
Washington State University
Tree Fruit Research &
Extension Center
Wenatchee, Washington

MATING DISRUPTION IS THE USE OF PHEROMONES to prevent the mating of a species by interrupting the normal process of mate location. Growers and field representatives are familiar with pheromones. These products are used in conjunction with traps to monitor the seasonal activity of several pests, for example, codling moth. First moth capture in pheromone traps is used to initiate the degree-day model predicting the development of codling moth.

Pheromones are chemicals produced by an insect that when released cause a specific response in other individuals of the same species. There are several different types of pheromones. For example, ants lay down a trail pheromone that is used by other members of the same species to locate a food source. Aphids release an alarm pheromone when disrupted by a predator or parasite that stimulates aphids nearby to move violently or even to drop from the plant. Sex pheromones are chemicals that, in the Lepidoptera, i.e. moths, are produced and released by females to attract males from long distance for the purpose of mating. A female releasing pheromone is said to be "calling" the male. The male flies upwind crisscrossing the pheromone plume and following the increasing concentration until it finds the source, the female *(see Figure 1a)*. After mating has occurred, the female ceases calling.

Pheromones of many different insects have been identified and synthesized. When small amounts of the pheromone of a species are incorporated into small rubber or plastic dispensers and placed in a trap, males of that species are attracted to the trap as if it were a calling female. The pheromone of most moths is comprised of from two to six components. In general, these chemicals are simple molecules, eight to 14 carbons long with some double bonds and an alcohol (OH) or acetate (COOH) group at one end.

The specificity of pheromones comes from the different components or the different ratios of the same components. For example, the pheromone of several leafroller species known as pests of tree fruit crops worldwide are given in Table 1. The shorthand notation of the chemical names, (Z)-11-14Ac [(Z)-11-tetradecenyl acetate], is a way of describing the shape, size, and atomic composition of the molecule. It is sufficient for this discussion that you know the chemical (Z)-11-14Ac is different biologically from (E)-11-14Ac, though both have exactly the same number of carbon, hydrogen, and oxygen atoms. The important thing to note is that several leafroller species have pheromones with the same components but those components occur in different ratios. It is the differences in the ratios of these components that make the pheromone an attractant for only one species.

MATING DISRUPTION

Mode of Action

There are several possible ways that mating disruption could operate. Figure 1b conceptually shows how mating disruption might work. Dispensers placed in the field might mimic a calling female, thus attracting the male to many "false" sources, or they might release so much pheromone that the background concentration is high enough to block normal communication.

An understanding of how mating disruption works is important because different mechanisms would suggest different approaches to the application of the pheromones in the field. There are five main

FIGURE 1

A) Normal mechanism of mate location used by Lepidoptera, where the female releases pheromone (a calling female) and the male orients his flight toward the source. B) Mating disruption concept; pheromone is released from dispensers that act as false sources or pheromone release is at such a high rate that the male is disoriented or unable to detect the plume of the calling female.

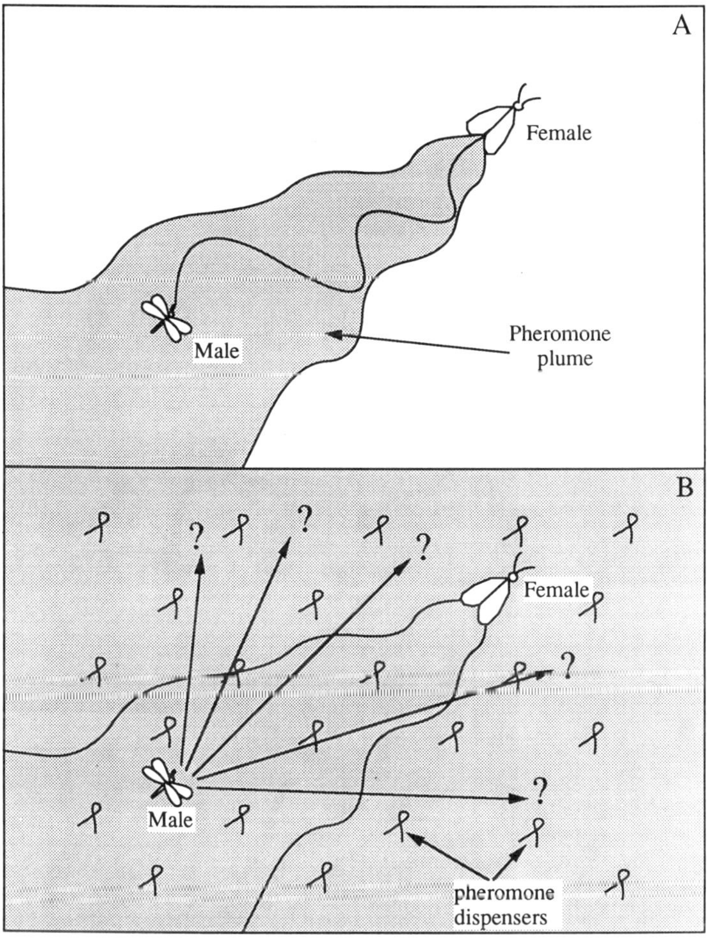

"modes of action" proposed by scientists working in this area to explain how mating disruption works and each is briefly discussed below.

PERIPHERAL AND CENTRAL NERVOUS SYSTEM EFFECTS: Continuous exposure to a stimulus can have an adverse effect on the sensory apparatus or nervous system of an organism so that it does not function normally. Sensory structures in moths (olfactory receptors) that detect pheromone are located on the antennae. When exposed to a constant source of stimulus (i.e., pheromone), output from these sensory organs declines rapidly, a condition referred to as adaptatin. Recovery of the sensory organis is rapid (2 to 3 seconds) once exposure to the stimulus is removed. High and uniform concentrations of pheromone in the field could effectively shut down the ability of the sensory organs to detect pheromone or at least the ability to orient to a source of the pheromone.

TABLE 1

Listing of several leafrollers known as pests of fruit crops throughout the world and the chemicals making up their pheromones.

Insect *(scientific name)*	Location	Pheromone	Percent/Ratio
Pandemis leafroller *(Pandemis pyrusana)*	Western N. America	(Z)-11-14Ac	96
		(Z)-9-14Ac	4
Threelined leafroller *(Pandemis limitata)*	Eastern NA & Canada	(Z)-11-14Ac	91
		(Z)-9-14Ac	9
Fruit tree leafroller *(Archips argyrospilus)*	N. America	(Z)-11-14Ac	15
		(E)-11-14Ac	10
		(E)-9-14Ac	1
		12Ac	50
Obliquebanded leafroller *(Choristoneura rosaceana)*	N. America	(Z)-11-14Ac	95
		(E)-11-14Ac	5
		(Z)-11-14OH	0.5-10
European leafroller *(Archips rosanus)*	Europe & N. America	(Z)-11-14Ac	100
		(Z)-11-14OH	10
		(E)-11-14Ac	0.1
Summer fruit tortrix *(Adoxophyes orana)*	Europe	(Z)-9-14Ac	9
		(Z)-11-14Ac	1
(Pandemis heperana)	Europe	(Z)-11-14Ac	85
		(Z)-9-14Ac	5
		(Z)-11-14OH	10

Exposure to high concentrations of a stimulus may result in the decline of a behavioral response lasting several minutes or even a few hours. The effect is evidently on the nerves and is referred to as habituation. In this situation, the nerve fires in response to a high concentration of the stimulus but does not recover in the normal manner. Thus, when exposed to a normal level of the stimulus, the nerve does not send the proper signal. This condition can occur even after the high concentration has been removed. Habituation caused by the exposure of moths to high concentrations of pheromone could play an important role in suppressing normal male responsiveness in mating disruption.

FALSE TRAILS. This concept assumes that the male moth can sense and respond to the pheromone stimulus, thus adaptation and habituation are of little or no importance. By placing numerous sources of pheromone in the field, male moths would spend time and energies following pheromone trails to a false source. If false sources were numerous enough, the likelihood of a male finding a calling female would be very low. If false trail following is an important mechanism in mating disruption, the ideal situation would incorporate an even distribution of numerous false pheromone sources all releasing about the same amount of pheromone as a calling female.

Most dispensers used in mating disruption release a high rate of pheromone relative to a calling female. Male moths will cease to be attracted to a pheromone source when the concentration of the pheromone gets too high, above a response threshold. However, false trail following may still occur even in the situation where a high rate of pheromone is released from a dispenser. The male moth may orient to a pheromone plume and follow the false trail until it reaches a concentration that is no longer attractive.

MASKING. This concept, like false trail following, assumes that the insect's sensory apparatus is working normally. In this situation, the background level of the pheromone is high and uniform enough to mask the odor trail from a calling female. Thus, the male's normal mate locating navigation system is useless, made ineffective by a "fog" of pheromone.

IMBALANCED SENSORY INPUT. Most Lepidoptera have a multi-component pheromone system. Even in situations where the same components are shared by different species, discrimination is achieved by differences in the ratios of the components. By releasing a single component of a pheromone system, or an altered ratio of the correct components, the atmospheric mixture might be so altered that males could not detect or orient to the balance of pheromone which would elicit a mate-seeking behavior. If males continually received signals out of balance with that to which their sensory apparatus is designed to detect, then mate-seeking behavior might be inhibited.

MODIFICATION OF RESPONSE. Antipheromone is the term given to chemicals that are active in inhibiting mate location. These chemicals may structurally resemble the true pheromone or may have a completely different chemical structure. It is thought that some antipheromones act by competing for the same receptor site on the sensory organ as the true pheromone. However, some antipheromones not structurally related to the true pheromone may block mate-finding in a completely different manner.

Pheromone Dispensers

The dispensers used to release a pheromone in mating disruption can be just as important as using the right chemicals. If the right pheromone is present in the dispenser, but it is released at too slow a rate, then insufficient quantities would be present in the atmosphere to block mate finding. If, on the other hand, the dispenser released pheromone too fast, it would be a waste of the chemical (thus expensive), and control would break down earlier than expected.

Several different types of dispensers have been used in mating disruption tests. Pheromones have been impregnated in rubber tubing, incorporated in plastic wafers, placed inside hollow fibers, and even formulated as sprays. Today, the most common dispenser systems used in mating disruption incorporate pheromones in plastic tubes, ampules, or packets that are designed to provide a slow release of the product over several months. These dispenser systems are placed in the field by hand at a rate of 150 to 400 per acre. Different dispenser systems containing the same pheromone may not provide the same level of pest control (mating disruption). Careful research is required on each type of dispenser system to determine how effective each is in a mating disruption program.

Benefits of Using Mating Disruption

The use of mating disruption to control a key pest, such as codling moth, would help to reduce the use of broad spectrum insecticides in Washington State orchards and thus increase the success of biological control. In 1989, apple growers applied an average of three insecticides to control codling moth. This accounted for 37 percent of all insecticides applied during that year. Secondary pests, such as aphids, leafhoppers, and leafminers, have developed resistance to insecticides used against codling moth. As a result, an average of three insecticides were applied for their control in 1989. Eliminating broad spectrum organophosphate insecticides used to control codling moth would allow better survival of natural enemies of aphids, leafhoppers, and leafminers. Eventually, these secondary pests would be controlled primarily by their natural enemies, eliminating the need for specific chemical controls on an annual basis.

The use of mating disruption over several years could result in the suppression of a pest population to very low levels. This might allow the cessation of controls for a period of time until pest populations again increased to dangerous levels. Just such a example has been reported in California where the Oriental fruit moth has been controlled by mating disruption for two or more years. The recovery of the pest population could be monitored with pheromone traps or fruit inspections at harvest.

Development of pests resistant to insecticides is an all too common occurrence in agriculture. The discovery that codling moth populations in California and Washington have developed high levels of tolerance to azinphosmethyl (4- to 7-fold) has caused concern among growers and field representatives. Alternative chemical controls for codling moth, synthetic pyrethroids, are anticipated to create as many or more problems than they solve. Mating disruption would provide an alternative control tactic that would help to reduce selection pressure for resistance development to chemicals currently registered on tree fruit crops. The conservation of chemical controls is an important consideration at a time when many are being lost to the re-registration process.

Farm workers are an important part of the tree fruit production system. The exposure of farm workers to insecticide residues during thinning and harvesting operations is an issue of increasing importance. Pheromones are essentially non-toxic to mammals, especially at the levels released in mating disruption programs. Since the chemicals are contained in the dispenser packet, no direct exposure to workers occurs.

Pheromones are not applied directly to the fruit and, as such, are not present as residues on the crop. The Environmental Protection Agency (EPA) has waived the requirement for a food tolerance on all pheromones registered to date. The use of mating disruption to control key pests could help reduce the already low levels of insecticide residues present on fruit. Consumers who have difficulty understanding the complex issues of food safety would welcome any technology that would reduce chemical residues on their food supply.

Constraints and Influencing Factors

Mating disruption will not work for every pest that has a pheromone communication system, nor will it work equally well in every location. Orchard size and isolation can be important factors in the success of mating disruption. Areas smaller than 10 acres, or even larger areas near a source of the pest, may doom mating disruption to failure. Since the pheromone treatment does not kill the pest, it is possible for individuals (mated females) to fly into the treated area and cause damage. For an insect such as the codling moth, isolation of between 110 and 220 yards from a source is thought to be adequate. It may be necessary to utilize border spraying as part of a mating disruption program if a source of the pest is nearby. More research is required to determine the size of a sprayed border necessary to provide protection.

Mating disruption is not considered to be a control technique that will reduce high pest populations to non-damaging levels. Mating disruption is intended as a tactic to keep populations at low levels. Thus, populations of the target pest must initially be at low levels or be reduced by other means, usually chemical controls. It may be necessary to combine mating disruption with chemical controls in the first year of implementation if the pest population is high.

The high specificity of mating disruption, control of only the target pest, can have a negative side. Pests that may have been kept at non-damaging levels by insecticide controls made against the target pest are going to be released from all but natural controls. For some of these pests, natural controls will be all that will be required to keep their populations below damaging levels. For others, the release from insecticide controls will mean that their population levels will increase, sometimes to damaging levels, and intervention with an insecticide may be necessary. Thus, the switch to mating disruption as a control for codling moth will not mean that all insecticides will no longer be needed in fruit production.

Mating disruption will result in the need for more intensive monitoring of orchards. Monitoring populations of pest and beneficial insects will be necessary to determine if insecticide intervention is required to control them. Monitoring insects has been encouraged in the past, but reliance on chemicals as the primary pest control tactic greatly reduced the necessity to put large amounts of effort into this activity. To take full advantage of the increased potential for biological control where mating disruption is used, monitoring efforts of both the pest and natural enemy populations will have to increase.

The cost of mating disruption will be higher, at least in the short term, than conventional chemical control programs. For example, the average chemical control program for codling moth, three applications of azinphosmethyl, costs about $36 per acre for materials and an additional $60 to $75 per acre for application expense. The cost of the mating disruption product registered for codling moth is $136 per acre with an application cost of about $20 per acre. Thus, the chemical control program would cost between $96 and $111 per acre, while the mating disruption program would be about $156 per acre. If, in the mating disruption program, some chemical controls for secondary pests, such as aphids and leafminer, could be eliminated because of increased biological control, then the cost differential between mating disruption and chemical control programs would be reduced. However, if pests that had been controlled by codling moth sprays increased as a problem, additional controls for them might have to be made.

Other factors might influence the success of mating disruption, and these require further research. The uniformity of the orchard canopy might influence the distribution of pheromone, leaving regions where control could fail. Orchards in windy areas or those on steep slopes might not be amenable to mating disruption because the concentration of pheromone throughout the orchard would not remain uniform or in high enough concentration to provide control. Weather patterns may have an influence on the success of mating disruption. Orchards in regions with high summer temperatures may require dispensers with different release rates than those in cooler regions.

Advantages in Washington State

There are several characteristics of fruit production in Washington State that make mating disruption an appealing technique and may contribute to its success. The insect pest complex associated with fruit production in Washington and other western states is smaller than that occurring in the East. The absence of pests, such as the apple maggot and plum curculio, should make mating disruption for codling moth and leafrollers economically more feasible. In addition, insect pests, especially the codling moth, are present at low levels in most commercial orchards due to chemical control programs. Thus, mating disruption could be directly implemented without the need of an aggressive pest reduction program.

Most orchards in Washington are not grown as isolated units but as contiguous plantings, often of several hundred acres. The large size of orchards, or contiguous orchard areas, and the tradition in Washington of implementing regional control programs for pests could increase the chances of success for mating disruption.

The habitat surrounding most orchards does not contain plants that act as alternate hosts for targets of mating disruption. Thus, the chance immigration of pests from these areas is greatly reduced compared with regions where orchards are surrounded by alternate hosts for target pests.

Natural enemies of several tree fruit pests have been studied in Washington. Without the disruption caused by chemical controls, these natural enemies have the potential to provide biological control of some pests. This is true for the western tentiform leafminer, *Phyllonorycter elmaella,* and possibly the apple aphid, *Aphis pomi* Much more research is required to fully understand the potential for biological control of some pests. However, there is sufficient evidence to indicate that a greater reliance on biological control can occur once insecticide use is reduced.

Regulation and Registration

The EPA regulates only certain uses of pheromones. Pheromones used in traps to monitor insect activity or to estimate population density are not a regulated use. However, if the intent is to prevent, destroy, repel, or otherwise mitigate any pest, then the use is regulated and the product must be registered as a pesticide under the provisions of the Federal Insecticide, Fungicide, and Rodenticide Act (FIFRA). Thus, if the intent is to control a pest, then the EPA considers that the product is a "pesticide" and regulates its use through registration.

The EPA has recognized that pheromones possess certain characteristics which make them inherently different from conventional pesticides. They have a unique (non-toxic) mode of action, are highly specific, occur naturally, and are used in very low volumes. The EPA classifies pheromones as semiochemical pesticides and has developed a tier-testing scheme to ensure that only the minimum amount of data is required to register them. Pheromones are not required to undergo more detailed testing if they successfully pass all of the tier-one data requirements. To date, all pheromones registered for use by the EPA have not triggered the need for testing beyond that required in tier-one. The EPA has also waived the requirement for tolerance on food crops since the chances of residues occurring are nil.

The EPA has registered pheromone products for control of the Oriental fruit moth, peach tree borer, and codling moth. Other non-food or non-fruit crop registrations include products for controlling the pink bollworm in cotton, gypsy moth in mixed forests, pine shoot borer, tobacco budworm, and boll weevil. The time required to register pheromones has been much less, one to two years, than that required for conventional pesticides. The EPA supports the concept of mating disruption as an environmentally safe pest control tactic.

Examples of Successes and Current Research

One of the most successful examples of the mating disruption technique is the control of Oriental fruit moth. Control of Oriental fruit moth with pheromone has been shown to be as good or better than conventional chemical control in several orchards in California. Since the registration of the pheromone in 1986, use on peach in Washington and California has grown to over 10,000 acres in 1990.

Research has been conducted for many years on the utility of mating disruption as a control for codling moth in Switzerland. In March 1991, a pheromone product received registration for the control of codling moth in the United States. Research in the United States has produced mixed results on the efficacy of this product, although under certain conditions it appears to provide very good control of codling moth. Research into factors that influence the control success of this product is continuing, and great care should be exercised if a decision is made to implement its use in any commercial orchard.

In 1990, a research project was initiated to examine the potential for using pheromones to control the pandemis leafroller in Washington orchards. The initial year's results were promising, and plans are to continue the research for two to three more years.

REFERENCES

Barrett, B.A., and J. F. Brunner. 1990. Types of parasite-induced mortality, host stage preferences, and sex ratios exhibited by *Pnigalio flavipes* (Hymenoptera: Eulophidae) utilizing *Phyllonorycter elmaella* (Lepidoptera: Gracillariidae) as a host. *Environ. Entomol.* 19: 803-807.

Bartell, R. J. 1982. Mechanisms of communication disruption by pheromone in the control of Lepidoptera: a review. *Physiological Entomol.* 7: 353-364.

Brunner, J. F. 1991. The future of tree fruit IPM. Wash. State Hort. Assoc. Proc. (in press).

Carroll, D. P. and S. C. Hoyt. 1984. Natural enemies and their effects on apple aphid colonies (*Aphis pomi* De Geer) (Homoptera: Aphididae) on young apple trees in central Washington. *Environ. Entomol.* 13: 469-481.

Charmillot, P.-J. 1990. Mating disruption technique to control codling moth in western Switzerland. pp. 165-182. *Behavior-Modifying Chemicals for Insect Management: Applications of Pheromones and Other Attractants.* [ed.] R. L. Ridgeway, R. M. Silverstein, and M. N. Inscoe, Marcel Dekker, Inc., New York.

Rice, R. E. and P. Kirsch. 1990. Mating disruption of Oriental fruit moth in the United States. pp. 193-211. *Behavior-Modifying Chemicals for Insect Management: Applications of Pheromones and Other Attractants.* [ed.] R. L. Ridgeway, R. M. Silverstein, and M. N. Inscoe, Marcel Dekker, Inc., New York.

Tinsworth, E. F. 1990. Regulation of pheromones and other semiochemicals in the United States pp. 569-603. *Behavior-Modifying Chemicals for Insect Management: Applications of Pheromones and Other Attractants.* [ed.] R. L. Ridgeway, R. M. Silverstein and M. N. Inscoe, Marcel Dekker, Inc., New York.

Willett, M. J. and J. F. Brunner. 1991. 1989 pesticide use survey in apple. Wash. State Hort. Assoc. Proc. (in press).

11

BENEFICIAL ARTHROPODS FOR PEAR PEST MANAGEMENT

HELMUT RIEDL
Mid-Columbia Agricultural
Research and Extension Center
Oregon State University
Hood River, Oregon

PEAR GROWERS ARE ALL TOO FAMILIAR WITH THOSE arthropods (insects and mites) which damage the tree, feed on foliage, or injure the fruit. Collectively, we call them pests. However, there are also many arthropods on pear which are beneficial. Perhaps the most widely recognized group of beneficials in pear orchards are pollinators, primarily bees. As we all know, their service is essential for the production of tree fruits. Therefore, to protect these, growers carefully avoid spraying harmful chemicals during bloom time.

A second and often neglected group of beneficials includes those arthropods (insects, mites, spiders) which help control pests. These natural antagonists (predators and parasitoids) can play an important role in the regulation of pest populations on pear. Most of us are aware of these "good bugs" in our orchards. Who has not heard of lady beetles or aphid lions and their voracious appetites for aphids? However, few of us have learned how to take advantage of the free biological control natural enemies provide and how to integrate it with chemical control practices. If we really want to make biological control work, we have to become as protective about natural enemies as about the pollinators. Growers must begin to realize that the service natural enemies can provide is also essential and can save them money in the long run. Many pear pests have effective natural enemies which contribute to various degrees to their control. If we want to make use of them, we must first recognize the different natural enemy species which we may find in pear orchards, and know their biologies, life cycles, feeding habits, and susceptibility to pesticides.

This chapter will focus on those beneficial arthropods which are important in the biological control of pear psylla. Chemical control of this important pear pest has always been difficult and is becoming more so because of serious resistance problems and a lack of new registrations. Therefore, pear growers will have to rely more and more on natural enemies to achieve acceptable control of pear psylla. Unfortunately, few publications exist which provide information and illustrations of the beneficial arthropods discussed here. A few sources are given following the references.

ARTHROPOD COMPLEX ON PEAR

In southern Oregon, about 230 different species of insects, mites, and spiders comprise the arthropod complex on pear. Only five percent of them can be considered serious pests and 40 percent are natural enemies. In pear orchards with standard spray programs, it is difficult to find natural enemies since they are very susceptible to broad spectrum pesticides (pyrethroids, Guthion, amitraz) and since pest populations are generally too low to support them. However, they can be found on unsprayed or abandoned pear trees or in orchards where selective pesticides are used. The natural enemy complex on pear consists of many generalist predators and parasitoids. The generalist predators may feed on spider mites, aphids, scale insects, and other plant-feeding arthropods. Almost all feed on pear psylla to various degrees.

Natural Enemy Complex of Pear Psylla

Wherever pear psylla occurs, it is attacked by a number of generalist predators including predaceous true bugs (Hemiptera), green and brown lacewings (Chrysopidae, Hemerobiidae), lady beetles (Coccinellidae), spiders, and others. In addition, a few fairly host-specific hymenopterous wasps in the family Encyrtidae parasitize pear psylla. An overview of natural enemy groups important for the biological control of pear psylla in western North America is given in Table 1.

TABLE 1

Natural enemy complex of pear psylla in western pear-growing areas.

Order	Family	Common name
Predators		
Hemiptera	Anthocoridae	Minute pirate bugs
	Miridae	Predaceous plant bugs
	Nabidae	Damsel bugs
	Pentatomidae	Stink bugs
Neuroptera	Chrysopidae	Green lacewings
	Hemerobiidae	Brown lacewings
	Raphidiidae	Snakeflies
Coleoptera	Coccinellidae	Lady beetles
Hymenoptera	Vespidae	Wasps
	Formicidae	Ants
Diptera	Syrphidae	Syrphid flies
Araneae		Spiders (many species)
Parasitoids		
Hymenoptera	Encyrtidae	Parasitic wasps

PREDACEOUS TRUE BUGS. Some of the most effective pear psylla predators belong to this group, including minute pirate bugs, predaceous plant bugs, damsel bugs, and stink bugs. All of these species are in the order Hemiptera. Biology and life cycle characteristics of predaceous true bugs are summarized in Table 2.

MINUTE PIRATE BUGS (ANTHOCORIDAE). This group of predators appears to be more closely tied to pear psylla than any of the other predator species. Several *Anthocoris* species have been described as pear psylla predators from western North America. The two most common native species are *A. antevolens* and *A. melanocerus*.

TABLE 2

Predatory Hemiptera.

	Minute pirate bugs (Anthocoridae)	Predaceous plant bugs (Miridae)	Damsel bugs (Nabidae)
Egg site	In leaf petiole, leaf vein	Under bark of young twigs	In stems of herbaceous plants
Overwintering site	Under bark scales, fallen leaves	Under bark of young twigs	In sheltered places
Overwintering stage	Adult	Egg, adult	Adult, egg
Generations	2-3	1-2	1

They overwinter as adults, often in groups, and disperse into orchards in the spring when temperatures rise above 50°F. They feed first on pear psylla eggs, but will also attack other insect or mite eggs if available. Their seasonal phenology is well synchronized with pear psylla. *A. nemoralis*, a European species, was introduced in 1963 in British Columbia and is now found in many parts of Washington. *Orius tristicolor* is smaller than the *Anthocoris* species. Adults of this species feed on pear psylla eggs and young nymphs. *Orius* is also an effective spider mite predator and can build up in the ground cover.

PREDACEOUS PLANT BUGS (MIRIDAE). The most important mirid predator of pear psylla is *Deraeocoris brevis*. In Oregon, it is generally more common on pear than anthocorids. This mirid overwinters as an adult in groups under bark scales. It is found on many host plants, but prefers arboreal habitat. On pear, it prefers to feed on pear psylla. It may be partially phytophagous. This would give *Deraeocoris brevis* the ability to sustain itself when prey is low. *Campylomma verbasci*, also known as the mullein bug, can be an important species in May when few other predators are present. It overwinters in the egg stage in bark tissue of young wood and has three to four generations per year. On apples, the mullein bug can damage the fruit, but it rarely causes injury to pears. Other mirids which can be locally important pear psylla predators are *Campyloneura virgula, Diaphnocoris provancheri*, and the ant mimic *Pilophorus sp.*

DAMSEL BUGS (NABIDAE). These are large predaceous bugs which usually do not occur in large numbers. Damsel bugs have one generation per year and overwinter as adults.

STINK BUGS (PENTATOMIDAE). Most species in this group are phytophagous. A few are also predaceous. One example is the sulcate stink bug, a common species in California pear orchards. However, other stink bugs may do serious damage to the fruit. Therefore, correct identification is important.

LACEWINGS. Green and brown lacewings are predaceous in the larval stage but less so in the adult stage. Adults are very dispersive insects and tend to migrate. Lacewings are often found in association with aphid colonies. On pear, lacewing larvae are important predators of pear psylla. Several green lacewings (Chrysopidae) have been identified from pear: *Chrysopa oculata* (British Columbia), *C. plorabunda* (California), and *C. carnea* (widely distributed). Eggs of green lacewings are characteristically supported on a stalk. In southern Oregon, *C. carnea* is considered the third most important predator of pear psylla after the mirid *D. brevis* and lady beetles. The brown lacewings (Hemerobiidae) glue their eggs directly on the plant substrate. In some areas and in some years, brown lacewings are more common than green lacewings. Lacewings are present in pear orchards from May until October, with a peak in July and August. *Hemerobius angustus* has been identified from California and

TABLE 3

Biology and characteristics of life stages of green and brown lacewings.

	Green lacewings (Chrysopidae)	Brown lacewings (Hemerobiidae)
Eggs	On stalks	Glued to substrate
Cocoon	White or yellowish, closed, not transparent	Loosely woven, transparent
Adults	Green or yellowish, no hairs	Brown or beige, hairy
Overwintering stage	Larva in cocoon, adult	Larva in cocoon, adult
Generations	3-4	3-4

H. pacificus from British Columbia. Up to six different hemerobiids were recently found on pear in northern Oregon. Biology and characteristics of life stages of green and brown lacewings are compared in Table 3.

SNAKEFLIES (RAPHIDIIDAE). This unique-looking insect has a long neck and can raise its head above the rest of its body. Snakeflies belong to the same order (Neuroptera) as the green and brown lacewings. The wings are transparent with many veins and are held rooflike over the abdomen, similar to the lacewings. The larvae inhabit galleries of bark beetles and other wood-feeding insects. They are predaceous. The adults feed on aphids, various other insects, and also on pear psylla. Snakeflies can be seen in pear orchards through most of the season. The larvae overwinter. Depending on the species, development may require from one to three years.

LADY BEETLES (COCCINELLIDAE). This is a large family of beetles (Coleoptera), most of which are predaceous on aphids, scale insects, and spider mites. The following species have been identified from pear in British Columbia: *Hippodomia quinquesignata, Adalia frigida, Calvia duodecemmaculata, Coccinella transversoguttata,* and *Coccinella convergens.* Thirteen different species were found in association with pear psylla in southern Oregon, among them: *Scymnus marginicoelis, Olla abdominalis, Cyloneda polita, Stethorus picipes,* and *Hippodamia convergens.* Some lady beetles such as the aphid-feeding *Adalia* species have one generation, others such as the scale-feeding coccinellids have from two to four generations depending on the climate. Adult coccinellids tend to migrate and congregate at overwintering sites. Lady beetles are not considered to be very effective pear psylla predators. However, in some years they can become quite numerous on pear and outnumber other species.

OTHER PREDATORS. Spiders are commonly found on unsprayed pear trees. However, their role as predators of pear psylla and other orchard pests is not well understood. In one study in northern Oregon, spiders from at least nine families were found in association with pear psylla. These included web spinning, jumping, and crab spiders.

Vespids such as the yellow jacket, *Vespula pennsylvanica,* and the bald-faced hornet, *V. maculata,* have been observed feeding on pear psylla. Wasps would be difficult to manage in an integrated control program. To protect pickers, wasp nests would have to be removed prior to harvest.

Ants are often found in association with honeydew-producing insects such as aphids and also pear psylla. A recent study demonstrated that, in addition to feeding on honeydew, certain species of ants also prey on pear psylla immatures. It is probably difficult to maintain ground-nesting ant colonies in irrigated orchards.

Several species of syrphid flies are known predators of pear psylla nymphs. Their numbers are usually too low to have a significant impact on psylla populations.

Parasitoids. Two encyrtid parasitoids attack pear psylla: *Trechnites insidiosus* and *Prionomitus mitratus*. These two species are quite host-specific and only parasitize pear psylla and other psyllids.

Trechnites insidiosus is the more common species. The female stings young pear psylla nymphs (1st to 3rd instar) with its ovipositor. The parasitoid larva does not develop until the pear psylla nymph reaches the 4th or 5th instar. The pear psylla nymph dies, and the parasitoid larva develops inside the shell of the dead nymph. This is called a mummy. The adult parasitoid emerges from the mummy about two weeks after pear psylla adults. This synchronizes the emergence of *Trechnites* adults with the next generation of pear psylla nymphs. The percentage of parasitized nymphs increases with each generation. In the fall, parasitized nymphs seek shelter under bark scales. *T. insidiosus* overwinters in the larval stage inside the mummy. Unfortunately, hyperparasitism reduces the effectiveness of *T. insidiosus*.

The other primary encyrtid parasitoid, *Prionomitus mitratus*, is larger than *T. insidiosus*, but is not as common. The species was introduced from Europe in 1963 and released in British Columbia.

Monitoring Natural Enemies

The beating tray is a widely used method for monitoring pear psylla adults, the immatures, and most adults of the predators discussed here, and adults of the two encyrtid parasitoids. The beating tray consists of a circular or square metal frame covered with a white cloth. It should have a surface area of 400 square inches. The tray is held horizontally under a branch. Insects are dislodged with three sharp taps to the branch, they fall on the cloth-covered tray and are identified and counted. For consistency, it is important to select branches of uniform size for tapping; take tray counts only during the cool part of the day and when there is no wind. Percent parasitism can be determined from counts of pear psylla nymphs and mummies on leaves.

Is it possible to predict from beating tray counts if predators will successfully control pear psylla? How many natural enemies are needed to regulate a pear psylla population and prevent unacceptable damage to the fruit? Unfortunately, there are no ready answers yet to these questions. Predator-prey ratios have been developed for interactions between a single predator and its prey. For instance, the likelihood of biological control of spider mites on apple in Michigan and on almonds in California can be predicted from the ratio of a phytoseiid predator to spider mites. However, when several natural enemy species are involved in regulating a population, as is the case with pear psylla, it becomes more difficult to predict the outcome of these interactions. Also, the fact that the natural enemy complex of pear psylla changes seasonally and from year to year adds to the difficulty to develop predictive information. Therefore, interactions between natural enemies and pear psylla need to be carefully monitored and supplemented with selective chemical control, if necessary, to prevent injury to the fruit.

Predator Efficiency or Which Predator Has the Greatest Appetite?
Predator species vary in their ability to respond to a building population of prey and regulate it. Several factors determine the outcome of predator-prey interactions. Among them are the searching ability, consumption rate, and reproductive rate of a predator species. The amount of prey consumed varies from species to species. Minute pirate bugs, predaceous plant bugs, and lacewings are all effective aphid predators. Table 4 shows the number of aphids consumed during nymphal or larval development of several European predator species. A small species such as *Orius* consumes less during its development than the larger predators, *Deraeocoris*, or a syrphid fly larva.

TABLE 4

Aphid consumption during larva/nymphal development of different predator species.

Predator species[1]	Common name	No. aphids
Anthocoris nemorum	Minute pirate bug	200
Orius sp.	Minute pirate bug	40-200
Deraeocoris ruber	Predaceous plant bug	200
Chrysopa carnea	Green lacewing	200-500
	Syrphid fly	400

[1]*Data compiled from several European studies.*

Little comparative information is available on the consumption of pear psylla eggs and nymphs by different predator species. In laboratory studies, the introduced minute pirate bug, *Anthocoris nemoralis*, preferred pear psylla nymphs, while the native green lacewing, *Chrysopa oculata*, preferably fed on eggs. *A. nemoralis* consumed a total of 17.8 immatures per day and *C. oculata* 22.4. More detailed information on prey consumption is available for the mirid predator *Deraeocoris brevis* (see Figure 1).

FIGURE 1

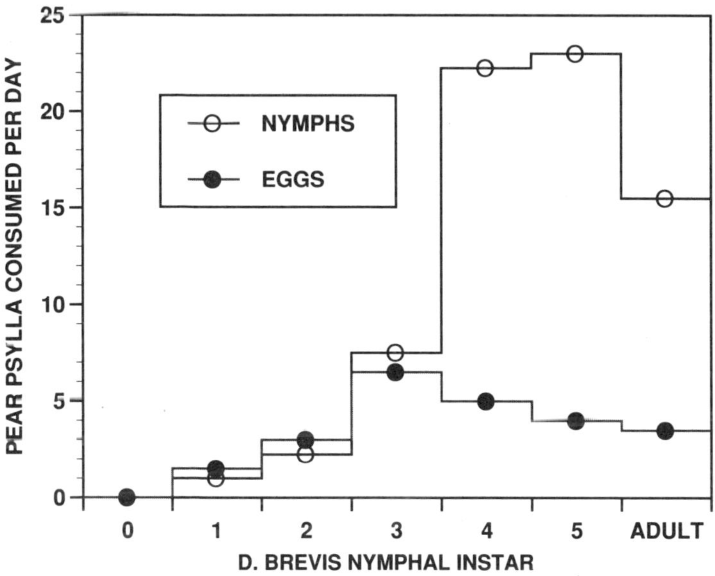

The first two instars of *D. brevis* prefer eggs to the nymphs of pear psylla, the third instar consumes eggs and nymphs in equal numbers, and the older *D. brevis* nymphs as well as adults feed primarily on pear psylla nymphs. *D. brevis* consumed about 400 pear psylla eggs and nymphs during development from first instar to the adult stage. The number of prey consumed per day or during the development of a predator is only one indication of predator efficiency. Little information is available on the ability of predators to search out prey and on their rate of reproduction.

Factors Contributing to the Successful Regulation of Pear Psylla Populations by Natural Enemies

Studies in Europe and North America have shown that whenever biological control of pear psylla develops, it is accomplished not by a single natural enemy species (the "silver bullet") but by the combined effect of a complex of different predatory species and a few parasitoids. For instance, 56 beneficial arthropods were associated with pear psylla, according to a recent survey of pear orchards in France.

Of those, only six were abundant natural enemies of pear psylla. About 35 species were identified as being predators of pear psylla in southern Oregon.

For successful seasonal biological control of pear psylla, it is important that the interaction between natural enemies and pear psylla begins early in the season and continues through the summer into fall. Late immigration of natural enemies is often the cause for failure since pear psylla has a high reproductive potential, and natural enemies, if they arrive too late, cannot catch up with a building pear psylla population before economic damage occurs. Results of a recent study in France serve as an illustration (*see Figure 2*).

FIGURE 2

Natural enemy complex of pear psylla in France.

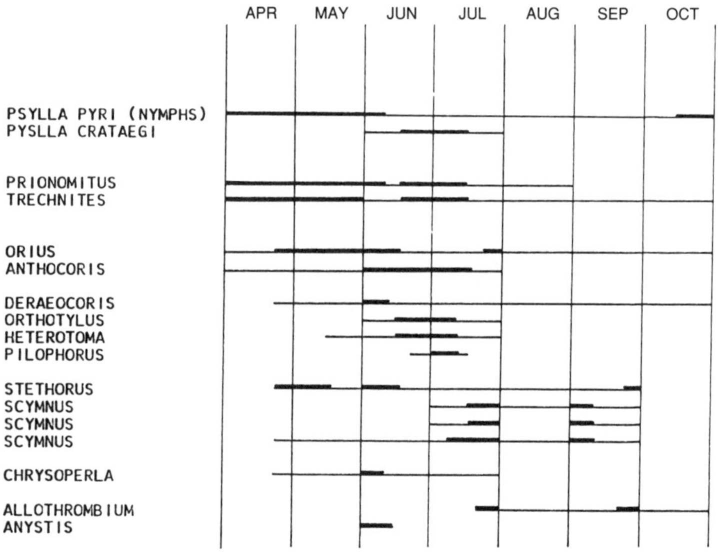

Regulation of pear psylla is provided by the combined activity of predators and parasitoids from April through October. The prey, *Psylla pyri*, a pear psyllid closely related to the species in our orchards, was present throughout the season with periods of greatest abundance until mid June and again in October when natural enemy activity declined. The two parasitoids *Trechnites* and *Prionomitus* were already active in April and provided early regulation of pear psylla. This was important since predator activity, except for some early anthocorids, was still low at that time. Predaceous mirids began to appear in larger numbers in June and July, but, except for a *Deraeocoris* species, were absent in late season. Several species of lady beetles (coccinellids) were primarily active during the second half of the season. A green lacewing species, *Chrysoperla*, was present in low numbers from May to July. Two predaceous mites provided some control, primarily in late season.

A within-season change in the natural enemy species complex was also documented in an unsprayed pear orchard in northern Oregon (*see Table 5*). In that location, predaceous bugs (primarily mirids) were the dominant predator group throughout the season. Their dominance increased from early to late season. Spiders, a green lacewing, and other natural enemies (parasitoids) were more abundant early on.

TABLE 5

Seasonal changes in pear psylla predators in an unsprayed pear orchard (northern Oregon: MCAREC, 1988).

		Percent of total	
Predator group	Early	Mid	Late
Hemipterans	39	75	87
Lacewings	9	5	2
Spiders	19	5	9
Other	33	15	2

From Booth (1991).

In addition to within-season changes in the natural enemy complex at a given location, there can also be considerable annual differences in the species complex (*see Table 6*). Observations in a pear orchard in southern Oregon over a three-year period showed a change from a complex dominated by coccinellid predators in 1964 to a complex dominated by chrysopid predators in 1965 and hemipterans in 1966.

Chances for success of biological control of pear psylla are enhanced if natural enemies reproduce and overwinter in an orchard. This would synchronize them with the appearance of pear psylla eggs and nymphs. Consequently, one would have to depend less on immigration of adult predators to initiate biological control of pear psylla early in the season.

The potential for biological control of pear psylla also depends on the location of a particular orchard site. The availability of most pear psylla predators varies with the habitat surrounding an orchard.

TABLE 6

Annual changes in pear psylla predators in an unsprayed pear orchard (southern Oregon: Medford).

	Percent			
Predator group	1964	1965	1966	Avg.
Chrysopids	24	71	25	39
Hemerobiids	2	5	4	4
Coccinellids	40	6	26	23
Hemipterans	34	18	45	34

From Westigard et al. (1968).

Predators are generally more plentiful in orchards close to natural habitat (e.g., woodland) or diverse agriculture and less plentiful if surrounded by other well-sprayed orchards. For instance, the mirid predator *D. brevis* is common on native vegetation such as ponderosa pine and black oak where it feeds on scale insects and aphids. Predators can disperse from these habitats into pear orchards and respond as pear psylla populations begin to increase.

Hedgerows along orchard borders may play an important role in the future as "insectaries" where predators can build up on alternate prey and then disperse to nearby orchards. For instance, in France, predators and parasitoids of pear psylla can maintain populations on nearby hawthorn trees where they attack *Psylla crataegi* *(see Figure 2)*. With the exception of the minute pirate bug, *Orius*, other pear psylla predators are rarely found in the orchard ground cover. However, flowering plants may be important as nectar sources for parasitic wasps which attack pear psylla.

Susceptibility of Natural Enemies to Pesticides

Why is biological control so unsuccessful in most pear orchards? The answer to this question is both simple and obvious—because of pesticides. Broad spectrum insecticides and miticides used in conventional control programs are very toxic to predators and parasitoids. Only a few (e.g., phytoseiid mites) have developed a degree of resistance to organophosphate and carbamate insecticides due to many years of exposure.

However, generalist predators and parasitic wasps are still quite susceptible to most pesticides used in today's conventional control programs on tree fruits. The International Organization for Biological Control (IOBC) has coordinated an intensive research effort in several European countries over the last 15 years to develop information on the toxicity of pesticides to various groups of beneficials. Some results of these evaluations are shown in Table 7. Insecticides are rated in terms of their toxicity to minute pirate bugs (Anthocoridae), lacewings, parasitic wasps, and spiders.

TABLE 7

Toxicity of selected insecticides to natural enemies.

	Rating[1]				
	Anthocorids		Lacewings	Parasitic	
Insecticide	Adult	Larva	Adult & Larva	Wasps	Spiders
Amitraz	1-2	1-3	1-2	2-4	4
Guthion	1-2	1-3	2-4	3-4	
Thiodan	2-4	3	2-3	2-3	2-3
Pounce	4	4	4	3-4	4
Zolone	1	1-2	1-3	1-3	2
Imidan	1-2	1-2	3	2-4	
Phosphamidon	1	2	2	3	1
Rotenone plus Pyrethrin	1-4	2-4	2-4	1-3	2
Dimilin	1	1		3	
Fenoxycarb	1	1	1	1	

From Staübli et al. (1984).

[1] *1) Harmless <25% mortality; 2) Moderately harmful 26-50% mortality; 3) Harmful 51-75% mortality; 4) Very harmful >75% mortality.*

For instance, amitraz, the widely used foliar material for pear psylla control, is not very harmful to anthocorids and lacewings, but will cause high mortality to parasitic wasps and spiders. Guthion, on the other hand, is relatively harmless to adult anthocorids, but is quite toxic to anthocorid nymphs, parasitic wasps, and other natural enemies. The pyrethroid Pounce (permethrin) is the most toxic of all the insecticides listed in Table 7. Other pyrethroids such as Asana (esfenvalerate), Baythroid (cyfluthrin), and Pydrin (fenvalerate) are probably as harmful to natural enemies as Pounce. All pyrethroids are very toxic to predatory mites. The organophosphate compounds Zolone (phosalone; registration cancelled on pears), Imidan (phosmet), and phosphamidon are rated as slightly to moderately harmful. The botanicals pyrethrin and rotenone have a brief broad spectrum knockdown activity but become selective due to short residual activity.

With some exceptions, the insect growth regulators Dimilin (diflubenzuron; not yet registered for pears in the United States) and fenoxycarb (not yet registered for pears in the United States) are essentially harmless to beneficials and cause no mortality. AgriMek (avermectin), a new miticide/insecticide, has broad spectrum activity when it is applied. However, surface residues are shortlived while subcuticular deposits are very persistent. Therefore, after an initial knockdown, avermectin has no contact activity and can be considered selective.

The toxicity of a pesticide depends on how it is applied to an organism. Topical application (e.g., applied as spray) causes considerably higher mortality than exposure to residue. In tests with the mirid predator *Deraeocoris brevis,* topical application of organophosphate insecticides and mineral oil resulted in very high mortality (*see Table* 8). However, the toxicity of one-day-old residue was much reduced. The only exception was Guthion which also had high residual toxicity to *Deraeocoris brevis* nymphs. Thiodan (endosulfan) had lower topical and residual toxicity to *Deraeocoris brevis* than other materials. Surprising was the high toxicity of one percent mineral oil.

TABLE 8

Toxicity of pesticides to *Deraeocoris brevis* nymphs.

Material	Rate ai lb./100 gal.	Percent mortality at 24 h Residue, 1 day	Topical
Guthion	0.5	83	100
Imidan	0.5	23	100
Zolone	0.75	6	94
Thiodan	0.5	6	31
Mineral oil	1 qt	0	57
	1 gal	67	100

From Westigard (1973).

Anthocorids are reportedly also quite susceptible to mineral oil applied during the prebloom period. Predaceous plant bugs (mirids) are probably more susceptible to prebloom sprays of sulfur and Morestan (oxythioquinox) than anthocorids.

POSSIBILITIES FOR BIOLOGICAL CONTROL UNDER CONVENTIONAL CONTROL PROGRAMS

What changes can be made to present conventional pest control programs on pear, so natural enemies can survive and contribute to the control of pear psylla and other pest species? Three suggestions can be made for modifying conventional programs and creating a healthier orchard environment for biological control. The first suggestion is to shift to insecticides and miticides which are less harmful to natural enemies. Table 6 can provide some guidance in the selection of materials with a more favorable spectrum of activity. Unfortunately, few pesticides which are used in conventional programs today offer a degree of selectivity. Among codling moth materials, Imidan has generally greater selectivity *(see Tables 6 and 7)* to generalist predators than Guthion but it has not been used widely because of higher cost and shorter residual activity. In spite of these shortcomings, a cover spray program with Imidan for codling moth control may be a preferred alternative, due to its greater selectivity to natural enemies.

Few alternatives exist to amitraz for summer control of pear psylla. More information needs to be developed on the selectivity of AgriMek at higher rates which are needed for foliar pear psylla control. Only Thiodan is available as a more selective alternative to pyrethroid insecticides for prebloom control of overwintering pear psylla adults. Biological spider mite control, which is so successful on apples, is not an effective option on pear because of the increased sensitivity of pear foliage to mite feeding, a low treatment threshold (1 to 2 feeding mites/leaf), and the use of disruptive pesticides in a seasonal spray program (pyrethroids, amitraz). The less disruptive miticides should be used, such as Vendex (fenbutatin oxide), unless organotin resistance is present, and AgriMek at lower rates.

Another way of making a conventional program more selective is to limit the use of disruptive pesticides to those times of the season when they least impact natural enemies and cause outbreaks of secondary pests. It is well-known that pyrethroids can have a severe impact on natural enemies and cause spider mite outbreaks when used after bloom. Therefore, pyrethroids should be limited to prebloom use on pears and should not be recommended during the foliar period. The pear industry has generally followed these recommendations for pyrethroids.

A certain degree of selectivity can also be achieved by lowering overall pesticide use in pear orchards. With intensive monitoring and a continuous review of the damage potential of pest populations, pesticide use can be kept to the essential minimum and possibly be reduced by 20 to 30 percent below present levels. Pesticide rates should be adjusted to the minimum necessary to suppress pest populations to subeconomic levels. Monitoring procedures and cultivar-specific treatment thresholds are available for the major pear pests. Properly used, monitoring programs permit a continuous evaluation of the risks of crop damage from various pests. Pesticides are applied when pests exceed levels which are likely to damage the crop. If pest populations stay below threshold levels, no control is required. This approach to pest control permits the presence of pest species in an orchard, albeit at levels below treatment thresholds. Judicious pesticide use and presence of low levels of prey will allow biological control to develop. In contrast, the traditional approach, still widely practiced today, seeks to eradicate and keep pests at levels where they are barely detectable. Lack of food (prey) is another reason why natural enemies are rarely found in most commercial pear orchards.

BIOLOGICAL CONTROL UNDER SELECTIVE PROGRAMS

Although conventional programs can be modified and improved to some extent with intensive monitoring, use of treatment thresholds, and cultivar-specific information on pest susceptibility, biological control will not become a major component of the overall pest control program on pear until broad spectrum insecticides and miticides are replaced by selective control tactics. The selective control program for pears was pioneered by Dr. P. H. Westigard at Oregon State University's Southern Oregon Experiment Station. A major objective of the selective program is to maximize the effect of biological control. This is accomplished by: a) utilizing minimally disruptive control tactics to achieve acceptable levels of pear pest control; b) minimizing pesticide applications by using treatment thresholds and intensive orchard pest monitoring; and c) modifying the use pattern of pesticides to make them less disruptive.

An essential feature of selective programs is the use of nondisruptive control tactics. While in the past, insecticides developed for codling moth were often broad spectrum with too little selectivity, some of the newer products have more desirable characteristics from an IPM perspective. For instance, nondisruptive control tactics for codling moth which have a fit in selective programs are the insect growth regulators Dimilin and fenoxycarb, use of the sex pheromone for mating disruption, and granulosis virus. Of these control tactics, only mating disruption is now fully registered for use on pears.

Mites can be selectively controlled with Apollo (clofentezine), Savey (hexythiaxzox), and selective rates of Vendex or AgriMek. Savey is registered on pears but not available for sale. AgriMek is close to registration. Selective prebloom control of pear psylla is achieved with mineral oil (plus sulfur if pear rust mite is a problem). Morestan (oxythioquinox), if needed, can be applied at the pink stage. The selectivity of oxythioquinox needs to be further evaluated. Diflubenzuron plus oil and fenoxycarb, if applied against codling moth, also provide selective pear psylla control during the summer. Insecticidal soap (Safer) sprays are not very effective but may be adequate as a supplement to biological control. One problem with soap sprays is their phytotoxicity to the fruit.

The implementation of the selective program has not progressed very much past the research stage in the United States, since several of the critical components of the selective program are still awaiting registration. The lack of a registered selective control method for codling moth has been a major road block to the implementation of this program. Selective programs have been more successful in Europe and are widely accepted by apple and pear growers in Italy, France, Germany, Switzerland, and the Netherlands. In large part, their success in Europe is due to the availability of selective pesticides, and a strong cooperative research and extension effort to develop and implement these programs.

One selective control program which is being adopted by a limited number of orchardists is the organic approach. Pesticides used in organic production must fit specific guidelines. These requirements restrict the pesticide choices available to organic fruit growers. Organic pesticides include botanicals, mineral and vegetable oils, sulfur, insecticidal soaps, and microbials. Generally, these materials are safer than synthetic pesticides and more selective, allowing biological control to develop. The reasons why most of these products have not found wider acceptance in conventional pear production is their lower efficacy and short residual activity, which require frequent application.

REFERENCES

Booth, S. R. 1991. The potential of indigenous natural enemies to regulate pear psylla in the Hood River Valley, Oregon. Ph.D. thesis, Oregon State University.

Burts, E. C. 1963. The pear psylla in central Washington. Wash. State Univ. Agric. Exp. Sta. Circ. 416:13 pp.

Gut, L. J., W. J. Liss, and P. H. Westigard. 1991. Arthropod community development and organization in pear. *Environmental Management* 15:83-104.

Herard, F. 1985. Analysis of parasite and predator populations observed in pear orchards infested by *Psylla pyri* (L.) (Hom.: Psyllidae) in France. *Agronomie* 5:773-778.

Herard, F. 1986. Annotated list of the entomophagous complex associated with pear psylla, *Psylla pyri* (L.) (Hom.: Psyllidae) in France. *Agronomie* 6:1-34.

Paulson, G. S. 1990. Evaluation of the possible role of predaceous ants in the management of pear psylla *Cacopsylla pyricola* (Foerster) (Homoptera: Psyllidae). Ph.D. thesis, Washington State University.

Riedl, H. and P. H. Westigard. 1988. Monitoring and economic thresholds

of major pear pests. Proceed. 102nd Ann. Mtg. Oregon Hort. Soc. 79:87-98.
Staübli, A., M. Hachler, P. Antonin, and C. Mittaz. 1984. Tests de novicite de divers pesticides envers les enemis naturels des principaux ravageurs de vergers de poiriers en Suisse romande. Revue suisse *Vitic. Arboric. Hortic.* 16:279-286.
Watson, T. K. and W. H. Wilde. 1963. Laboratory and field observations on two predators of pear psylla in British Columbia. *Can. Entomol.* 95: 435-438.
Westigard, P. H., L. G. Gentner, and D. W. Berry. 1968. Present status of biological control of the pear psylla in southern Oregon. *J. Econ. Entomol.* 61:740-743.
Westigard, P. H. 1973. The biology of and effect of pesticides on *Deraeocoris brevis piceatus* (Heteroptera: Miridae). *Can. Entomol.* 105:1105-1111.
Westigard, P. H., P. B. Lombard, and D. W. Berry. 1979. Integrated pest management of insects and mites attacking pears in southern Oregon. Oregon State Univ. Agric. Exp. Sta. Bull. 634:41 pp.
Westigard, P. H., L. J. Gut, and W. L. Liss. 1986. Selective control program for the pear pest complex in southern Oregon. *J. Econ. Entomol.* 79:250-257.
Westigard, P. H. and P. D. Van Buskirk. 1988. A selective program for the pear pest complex in southern Oregon. Proceed. 102nd Ann. Mtg. Oregon Hort. Soc. 79:97-99.
Zwick, R. W. and G. J. Fields. 1977. Integrated control of pear psylla in the Hood River Valley. Oregon State Univ. Agric. Exp. Sta. Circ. 660:8 pp.

The following publications provide information as well as illustrations of natural enemies on pear:

1) *Pear Pest Management Handbook*, Univ. of Calif. Public. 4086 (1978). This handbook contains a well-illustrated section on beneficial insects and mites. A new pest management guide for apples and pears is due out in late spring of 1991. Contact UC IPM Manual Group, Univ. of California, Davis, CA 94.
2) *Nutzlinge in Apfelanlagen* (beneficials in apple orchards), Int. Organization for Biol. Control (IOBC). Vol. 3:242 pp., Pudoc, Wageningen, the Netherlands. This volume is a joint effort of several tree fruit entomologists in Europe and is by far the most comprehensive publication on beneficials in deciduous orchards. Good illustrations, keys, and biological information on many species of natural enemies. Unfortunately, this volume was published only in French and German.
3) *Tree Fruit Insects* (1981), North Central Regional Extension Publ. 63:66 pp. Written by J. F. Brunner et al. This publication has a short section (with illustrations) on beneficial insects and mites. This publication is now out of print but may still be available through some extension offices.

12

INSECT GROWTH REGULATORS

JAMES L. KRYSAN
United States Department of Agriculture
Agricultural Research Station
Fruit and Vegetable Insects Research Unit
Yakima, Washington

INSECT GROWTH REGULATORS (IGRS) ARE INSECT control materials that differ in chemistry and origin from most pesticides currently in use in the tree fruit industry. The IGRs bear many desirable attributes.

For tree fruit pest management in the United States, IGRs are still in the future because none as yet has reached full registration. If IGRs were ready for implementation, entomologists such as Drs. Everett Burts or Jay Brunner of the Washington State University Tree Fruit Research Center or Dr. Pete Westigard of Oregon State University would be a much better choice to write this chapter, because they have been studying IGRs in terms of the development of spray programs. Rather than address spray programs, I will provide a basic background on the origin of IGRs, how they differ from present and past pest control chemicals, and what makes them attractive. And finally, I will illustrate the potential of these materials by discussing how one class of IGRs is likely to be useful in the control of pear psylla.

IGRs have caught the fancy of entomologists because several features make them especially attractive as replacements for broad spectrum insecticides. Compared with broad spectrum insecticides, IGRs are more selective. Hence, they are environmentally safer and more compatible with integrated management schemes that include biological control agents. Additionally, insects rapidly develop resistance to current insecticides. Pesticides that are used with great frequency and are singularly and directly toxic to a broad array of organisms will eventually be lost to resistance, environmental concerns notwithstanding. IGRs represent entirely new classes of compounds, and intelligent use of these new classes should lessen the likelihood that resistance will develop.

Categories into which insect control methods can be placed include cultural, genetic, plant resistance, biological, and chemical. IGRs are among the chemically-based control approaches, a category that is likely to be critical for tree fruit insect management for a long time to come.

From the viewpoint of insect physiology, manipulating insects using chemicals makes good sense because living things are a coherent, delicately suspended mix of structural and dissolved chemicals that are held in a steady state—i.e., alive—with a continual input of energy and nutrients in chemical form. So, when you have a chemical system and you want to affect it, a direct way is with chemicals. In a sense, you are talking to a living thing in a language it can understand. IGRs offer a way to do that very selectively.

That selectivity stems from fundamental differences in how growth regulators act on insects compared with the action of most conventional insecticides. Based on the part of the insect's biology affected, virtually all chemicals now used to control insects can be assigned to one of three categories: neurotoxins, growth regulators, and behavior modifiers. The overwhelming majority of chemicals now used to control insects are neurotoxins.

Incidentally, it has been argued that the primary reason most insecticides are neurotoxins stems from their origins and how they were discovered. The organophosphorus insecticides were derived from the nerve gases that were first exploited for military purposes. Other insecticides were discovered by selection from large series of previously synthesized chemicals using simple bioassays that kill very quickly. About the only thing that kills quickly is a neurotoxin, so that's what got discovered. In the discovery and development processes, the real efforts were on chemistry. The biology was rudimentary in the extreme.

The origin of IGRs and behavior modifiers is entirely different. Their discovery is based on a thorough knowledge of how insects grow, develop, function, and behave. They have been discovered in two ways. First, given that normal processes are well understood, an insect can be exposed to selected chemicals and carefully observed for abnormalities to occur in development, behavior, or function. Chemicals that produce desired effects can be further developed. A second way involves the discovery and characterization of processes in insects that include, for example, hormones. By their nature, hormones act in very low concentrations to regulate and control particular biological processes with rather precise timing. They are secreted by one tissue and are carried via the blood to the tissue(s) on which they act. Over a period of several decades, insect hormones were discovered and biologically and chemically characterized; they then were used as models for the design and synthesis of chemical analogs that will interfere with normal insect growth and development.

Chemicals discovered by these two approaches differ tremendously in selectivity. All animals share basically the same neurochemical systems, due to a common evolutionary origin. Thus, with some minor differences due to secondary effects, neurotoxins are poisons for all animals. The IGRs act on functional systems unique to insects, or systems which insects share with, at most, close relatives; the materials are unlikely to affect organisms in other groups (*see Table 1*). Note that methyl parathion and malathion are both neurotoxins that are highly toxic to insects; the drastic difference in mammalian toxicity of methyl parathion and malathion is due to the presence of an enzyme in mammals that rapidly catalyses malathion to non-toxic metabolites. The IGRs induce debilitating changes in insects in the same dose range for which methyl parathion is toxic to mammals.

TABLE 1

Mammalian toxicities of some Insecticides and IGRs.

Compounds	Oral LD_{50} (mg/kg)
Methyl Parathion	1.1
Malathion	1,000
Diflubenzuron	4,640
Methoprene	34,500

Behavior-affecting chemicals are discovered in much the same way as IGRs, but many tend to be even more specific. Pheromones, for example, are behavior-modifying chemicals that aid the sexes of a single species to find each other, so that effort is not wasted chasing about after mates of a different species. Among closely related species, they are ultimately selective because they contribute to reproductive isolation between species. I cite behavior-modifying chemicals here as insect control agents of an order specificity beyond IGRs.

BRIEF INSECT ENDOCRINOLOGY

Arthropods, the group including insects, wear their skeletons on the outside; for growth, a new skeleton (made up of cuticle) must be formed inside the old, the old skeleton shed, and the new one swelled to the new, larger size and hardened. The process is called molting. Larval molts accommodate growth. The changes from larval characteristics to the adult form (a process called metamorphosis) also take place during molting. Molting and metamorphosis have fascinated physiologically-oriented entomologists for generations. Early entomologists showed with simple, but elegant and well-crafted experiments that hormones from the brain, corpora allata, and prothoracic gland controlled the various phases of molting.

The epidermis (which forms the cuticle) is the critical target tissue for hormones controlling molting and metamorphosis. One can arbitrarily start the process at the brain where prothoracicotropic hormone is produced and released; it stimulates the prothoracic gland to produce ecdysone. During any larval stage, the ecdysone acts first to program the epidermal cells and then to stimulate the actual molt. The corpora allata produce and release juvenile hormone *(see Figure 1)*, which acts on the epidermis to determine whether a molt produces another larva, a pupa, or an adult. A high hemolymph (=blood) titer for juvenile hormone at the programming time determines that a molt will yield another larva. A moderate titer yields a pupa, and, in the absence of juvenile hormone, an adult is produced.

Juvenile hormone was also found to play a hormonal role in many processes bearing no connection with metamorphosis *(see Table 2)*.

FIGURE 1

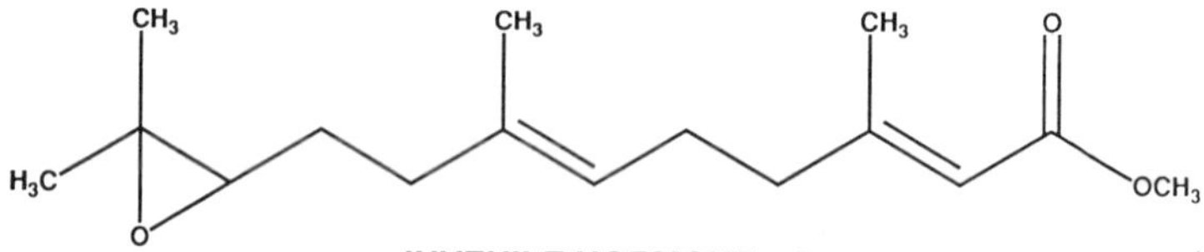

JUVENILE HORMONE – 3

TABLE 2

Insect processes regulated by juvenile hormone

Embryogenesis	Migration/Dispersal
Molting/Metamorphosis	Reproduction
Diapause	Communication
Phase Transformation	Pigmentation

INSECT GROWTH REGULATORS

Studies done over the first 70 years of this century provided a profound fundamental understanding of basic physiological processes, which lead ultimately to the discovery of IGRs potentially useful in insect control. Ecdysone and prothoracicotropic hormone have not served as useful models for designing IGRs because of their natures; the former is a steroid, the latter is a large and complex polypeptide.

Three types of IGRs are generally recognized—chitin synthesis inhibitors, juvenile hormone analogs and mimics, and anti-juvenile hormone agents. In addition, several materials that have other actions have been discovered but have not been given serious consideration for development as IGRs. Also, the anti-juvenile hormone agents, the most common group being the precocenes, have many appealing features and have been studied extensively.

Anti-Juvenile Hormones

As the name implies, and viewed simply, anti-juvenile hormones cancel the effect of juvenile hormone by blocking juvenile hormone synthesis in the corpora allata. An early instar treated with precocene molts prematurely into a diminutive, nonfunctional adult. They also sterilize some insects. Precocenes are extracted from plants, and it is thought they have evolved to protect plants from insects. A fatal disadvantage of the precocenes, from an economic perspective, is that they are so selective that none has generated sufficient interest to stimulate product development. The remaining two categories of IGRs include materials actively being developed for insect control purposes.

Chitin Synthesis Inhibitors

The chitin synthesis inhibitors are benzoylphenyl ureas. As the name indicates, they inhibit the synthesis of chitin, a complex carbohydrate that gives structure to the skeleton of insects and other arthropods. When treated with one of these compounds, an insect generally grows normally until the time to molt. Then, abnormalities occur, which lead to death. Sometimes death follows quickly, but often it occurs far in the future. In addition to molting disruption, the benzoylphenyl ureas also interfere with feeding, act as ovicides, and cause delayed mortality. These latter organismic effects probably stem from the molecular event of chitin synthesis inhibition. Diflubenzuron (*see Figure* 2) is probably the best known compound in this group.

FIGURE 2

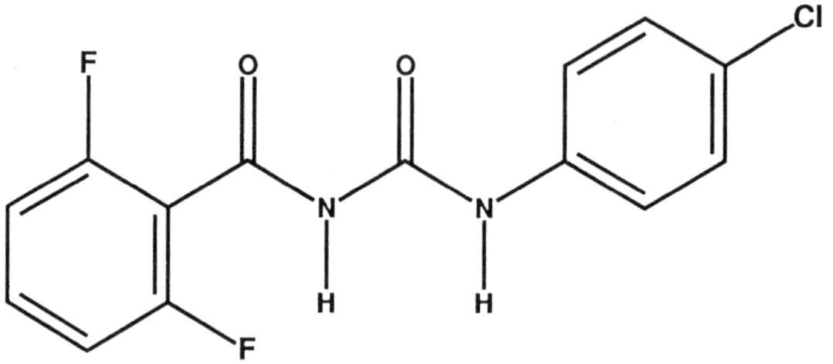

DIMILIN/DIFLUBENZURON TH 60-40 LOG MEAN

Juvenile Hormone Mimics and Analogs

The second group of IGRs includes the juvenile hormone analogs and mimics. When applied to an insect, they provide an abnormal source of juvenilizing agent with striking consequences. If the normal course of events calls for a molt to, say, the pupal stage, an abnormally high level of juvenilizing agent will result in another larval stage, a so-called supernumerary larval stage. Other physiological modes of action of juvenoid IGRs include embryocidal, ovicidal, sterilization, diapause disruption, and behavioral disruption. Theoretically, all the insect systems known to be influenced by juvenile hormone (*see Table* 2) are potential targets for action of a juvenoid IGR.

Juvenile hormone itself is relatively unstable and therefore not suitable for use as an IGR. The early juvenoid IGRs were true analogs of juvenile hormone and suffered from instability when exposed to ultraviolet light. This weakness seriously limits use in plant protection. However, another group of juvenoid IGRs, which some call juvenile hormone mimics, was discovered. While using highly sensitive bioassays to identify juvenile hormone, entomologists discovered that extracts of many plant tissues have juvenilizing properties. Identification of the causative factors revealed new chemical structures different from juvenile hormone (*see Figure* 3) and with much greater environmental stability. They have been used as models to design some highly effective juvenile hormone mimics, two of which are being evaluated for control of tree fruit pests.

FIGURE 3

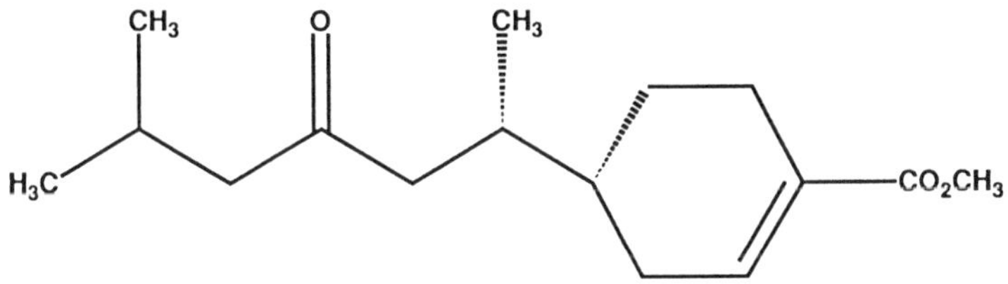

JUVABIONE (from balsam fir)

PEAR PSYLLA AND JUVENOID IGRS

The pear system is an excellent candidate for use of IGRs because their selectivity tends to preserve the natural enemies that can contribute to the control of pear psylla. It is especially important that psylla be controlled by an integrated system including several control factors because psylla rapidly gains resistance to pesticides.

Studies to date have shown that juvenoid IGRs can induce abnormalities in three different physiological systems in psylla: reproduction, morphogenesis, and diapause.

Westigard demonstrated that formulations of the earliest of the juvenoid IGRs (methoprene and hydroprene) caused psylla to lay sterile eggs. Those earliest juvenoids are not stable to ultraviolet light, so they are not practical for most plant protection applications. More recently, McMullen reported that the juvenoids pyriproxyfen and fenoxycarb (*see Figure 4*), when applied to psylla female adults, cause

FIGURE 4

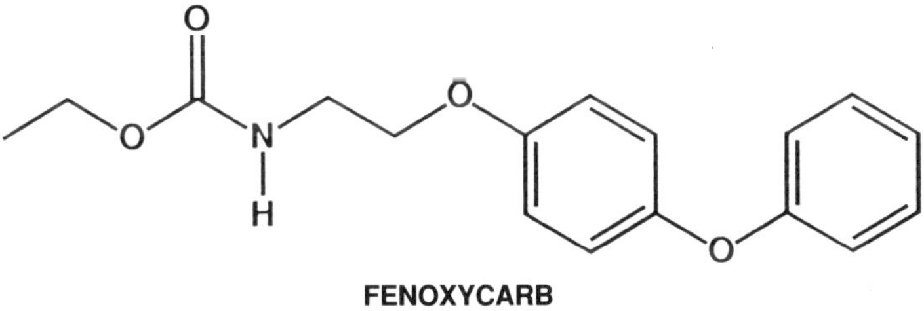

FENOXYCARB

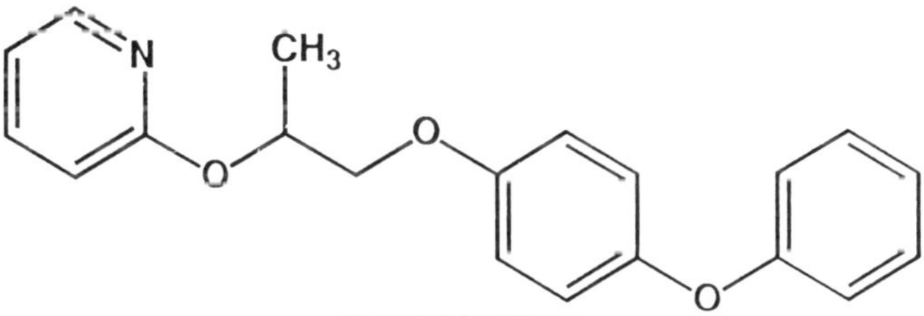

PYRIPROXYFEN

them to lay sterile eggs. Additionally, direct treatment of eggs prevented hatch; younger eggs were more susceptible, while eggs within a few days of hatching were not susceptible. It is difficult to determine whether the "sterility" effect is the result of direct action on the female, or on the egg while it is in the female. In recent studies, we observed that the sterility caused by pyriproxyfen disappeared within five to six days after removal of the female from the treated pear substrate, which might mean that the egg is affected directly. Dr. Everett Burts has suggested that the sterilization/ovicidal effect may be useful against winter forms in spring.

European workers found fenoxycarb to cause morphogenetic defects in the pear psyllid *Cacopsylla pyri*. McMullen found that nymphs did not survive treatment with juvenoids; with pyriproxyfen, almost all the mortality of nymphs occurred in the molt from 5th instar to adult, whereas with fenoxycarb, substantial mortality occurred at earlier molts. The prospect of attacking psylla nymphs merits further consideration.

Premature termination of diapause is the third possible mode of action for juvenoids against psylla. Discussion of this effect requires preliminary comments on the process of diapause.

Diapause

To adapt an insect for survival through an inimical season, processes have evolved which generally entail the insect having two qualities:

1) A specific stage with features expressly for survival through the hostile season. Among those features is the ability to be dormant, so it is often called the dormant stage.

2) A mechanism to ensure that the insect is in the dormant stage before the hostile season arrives. For example, if the hostile season is a cold winter, then the insect must be in the winter-hardy stage before cold weather arrives. Diapause is the name of that internal physiological process that imposes dormancy of the insect when the environment would favor growth and development. Once the environment becomes "cold", i.e., is below the temperature threshold for growth and development, the cold temperature enforces dormancy (since insects are cold-blooded animals). Quiescence is the term for dormancy that is imposed by the environment. The role of diapause is to couple the insect to the seasonality of its habitat.

Psylla nymphs are monophagous, feed on foliage, and are not adapted to survive a season when pear trees are leafless. Psylla adults that eclose in late summer are in reproductive diapause characterized by undeveloped ovaries, ample fat stores, lack of mating, and a morphotype (the winter form) that is easily distinguished from summer generation individuals. The reproductive diapause in pear psylla prevents oviposition until midwinter in the Pacific Northwest; diapause then terminates and the cold midwinter temperatures maintain quiescence until spring. When eggs hatch, foliage is present.

Bowers and Blickenstaff found that juvenile hormone terminated diapause in an insect, and suggested that insects might be controlled by the untimely termination of diapause. No pest control systems appeared however, mainly because laboratory tests showed that high doses of juvenoid IGRs were needed and the effect was rather quickly reversed. In sharp contrast, we have shown that the diapause of pear psylla is unusually susceptible to irreversible termination by exceedingly small doses of juvenoids, particularly fenoxycarb. In the laboratory and in cages in the field, we found that foliar applications of fenoxycarb terminated diapause in psylla placed on the foliage. Fenoxycarb treatment of an uncaged orchard block in autumn resulted in termination of diapause of psylla in the treated block, and post-diapause psylla appeared in adjacent, untreated rows.

These results suggested to us that psylla might be controlled via premature termination of diapause by treating pear in autumn with a juvenoid IGR such as fenoxycarb. It is well entrenched in entomological theory that diapause confers fitness for overwintering, so if diapause is terminated before winter, fitness should be reduced and, hence, less offspring produced the following spring. It has been rather simple to demonstrate that treatment with fenoxycarb in autumn causes ovarian development, mating, and oviposition before winter. But it is much more difficult to show that premature termination of diapause actually reduces overwintering fitness. A key complication for experiments is the dispersal habit of winter-form psylla in fall and spring; many psylla disappear from pear in fall and reappear in the spring, which complicates season-long experiments in the open orchard. We have demonstrated that diapause can be terminated, irreversibly, in uncaged field plots.

We have attempted to quantitatively monitor overwintering survival and post-winter survival and fecundity by placing winter form psylla in cages covering treated and control tree branches. Such cages were installed at various times in autumn. On several dates in late winter and spring, the cage-branch units were cut down and survival determined for adults and for eggs present on shoots from the caged branches. The latter was determined by excising egg-bearing shoots and placing their cut ends in water. Additionally, surviving females were placed on pear seedlings, and the number and viability of eggs was determined.

Eggs laid on the shoots in autumn did not survive; survival of eggs laid in the spring was significantly lower on treated branches compared with control branches. However, we were unable to demonstrate a significant difference in fecundity between treatments when females which had spent the winter in the cages were placed on seedling pear in the laboratory and viability was determined on the eggs laid on those seedlings. In the first winter of observations, survival of adults was the same in treated and control cages. However, in a second winter, the adults from the treated cage died sooner. We continue our efforts to assess the effects of premature termination of diapause on the overwintering fitness of psylla and look to the time that an experimental use permit is available to allow a strictly empirical test of the concept.

The approach has some attractive features. Successful psylla management depends on reducing the size of the first generation in the spring, so the target generation is correct. The method would help preserve natural enemies that are so important to psylla control. A critical complication is the dispersal habit of psylla; area-wide approaches would be needed, but the minimal size of the area is unknown.

CONCLUSION

The IGRs now available are products of detailed, long-term, basic studies done largely on insect species that are good "laboratory animals." From these intense studies, a solid body of general knowledge emerged, and it provides the the basis for designing chemicals that can cause sometimes subtle physiological, behavioral, or morphological effects on a great variety of insects. When these effects reduce fitness in the insect, they can be exploited for control of the insect.

Just as detailed knowledge of model insects was essential to develop the principles, so will there be need for a deep appropriate knowledge of the target insects. Subtle effects of IGRs which will ultimately contribute to the control of the target organismic pest must be recognized and exploited by incorporating into truly integrated management schemes.

REFERENCES

Bowers, W. S. 1990. Prospects for use of insect growth regulators in agriculture. M. Hoshi and O. Yamashita (eds.), *Advances in Invertebrate Reproduction 5*. Elsevier, Amsterdam.

Bowers, W. S. and C. C. Blickenstaff. 1966. Hormonal termination of diapause in the alfalfa weevil. *Science* 154:1673-1674.

Brunner, J. F. 1988. Insect growth regulators promising in pest control. *The Goodfruit Grower*, 39(5):33-38.

Burts, E. C. 1983. Effectiveness of a soft-pesticide program on pear pests. *J. Econ. Entomol.* 76:936-941.

Frischknecht, M., W. Jucker, M. Baggiolini, and A. Schmid. 1978. Mode of action and practical possibilities of an insect growth regulator with juvenile hormone activity in pear psyllid control.

Krysan, J. L. 1990a. Premature termination of reproductive diapause by a juvenile hormone mimic; a possible control strategy for pear psylla, pp. 387-391. M. Hoshi and O. Yamashita (eds.), *Advances in Invertebrate Reproduction 5*. Elsevier, Amsterdam.

Krysan, J. L. 1990b. Fenoxycarb and diapause: a possible method of control for pear psylla (Homoptera: Psyllidae). *J. Econ. Entomol.* 83:293-299.

Krysan, J. L. and B. S. Higbee. 1990. Seasonality of mating and ovarian development in overwintering *Cacopsylla pyricola* (Homoptera: Psyllidae). *Environ. Entomol.* 19:544-550.

McMullen, R. D. 1990. A report on pyriproxypfen and fenoxycarb for control of pear psylla *pyricola* Foerster, pp. 399-402. M. Hoshi and O. Yamashita (eds.), *Advances in Invertebrate Reproduction 5*. Elsevier, Amsterdam.

Oberlander, H. 1985. Hormone action during insect development, pp. 507-534. In M. S. Blum (ed.), *Fundamentals of Insect Physiology*. Wiley, NY.

Solomon, M. G. and J. D. Fitzgerald. 1990. Fenoxycarb: A selective insecticide for inclusion in integrated pest management systems for pear in the UK. *J. Hortic. Sci.* 65:535-540.

Retnakaran, A., J. Granett, and T. Ennis. 1985. Insect growth regulators. G. A. Kerkut and L. J. Gilbert (eds.), *Comprehensive Insect Physiology, Biochemistry and Pharmacology*, 12:529-601.

Solomon, M. G., J. E. Cranham, M. A. Easterbrook, and J. D. Fitzgerald. 1989. Control of the pear psyllid *Cacopsylla pyricola* in southeast England by predators and pesticides. *Crop Protection* 8:197-205.

Westigard, P. H. 1974. Control of the pear psylla with insect growth regulators and preliminary effects on some non-target species. *Environ. Entomol.* 3:256-268.

Westigard, P. H., L. J. Gut, and W. J. Liss. 1986. Selective control program for the pear pest complex in southern Oregon. *J. Econ. Entomol.* 79:250-257.

13

PESTICIDE RESISTANCE: PRINCIPLES AND APPLICATION

T. J. DENNEHY and
T. E. MARTINSON
New York State Agricultural
Experiment Station
Cornell University
Geneva, New York

THOUGH IT HAS BEEN OVER 80 YEARS SINCE THE first major account was published of an agricultural pest becoming resistant to a pesticide, it was not until the widespread development of insect resistance to DDT in the 1950's that most growers became familiar with the term resistance as it applies to pests. Since the 1950's, growers have learned to anticipate the eventual loss of effectiveness of pesticides, especially insecticides and fungicides, due to the development of resistance. Insects and mites plus plant pathogens account for the greatest number of resistance cases occurring in the world. However, important resistance problems have also occurred with other organisms.

Resistant bacteria, for example, pose a very serious threat of infection to hospital patients. Resistant Sporozoa such as the malaria pathogen are a serious threat to public health in the Third World and developing countries. Rodents possessing resistance to rodenticides pose serious public health problems in the largest metropolitan areas of the United States and elsewhere in the world. Cases of resistance to specific pesticides have been documented in frog, fish, crustacean, nematode, and weed populations. We conclude that essentially any organism that is controlled with chemicals possesses the potential to become resistant to the toxicant. Nonetheless, in terms of the number of resistance problems and their practical importance, it is the approximately 450 resistant species of insects and mites and 100 resistant species of plant pathogens that have caused humankind the greatest problem.

The objective of this chapter is to make resistance development and resistance management understandable. We will first outline the principles of what resistance is, how it develops, and the principle factors that affect the development of resistance. Then, we will outline resistance management strategies and give specific examples of the building blocks of a practical resistance management program.

We acknowledge that many of the concepts detailed herein are derived from the writings of two founders of pesticide resistance management—G. P. Georghiou of the University of California-Riverside, and the late R. M. Sawicki of the Rothamsted Experimental Station in Harpenden, England. We also recognize J. P. Nyrop, as well as other cooperators from Cornell University and elsewhere who have contributed to the research on resistance management.

WHAT IS RESISTANCE AND HOW DOES IT DEVELOP?

Resistance "develops" through the process of selection by a chemical on variability in susceptibility that exists within a target population of organisms. How this happens is easy to understand. Early on in the process of resistance development, only an extremely small proportion of individuals in a population may possess the genetic potential to survive exposure to the pesticide. However, every time the pesticide is applied, these "resistant" individuals survive in relatively greater proportions than do the susceptible types. In other words, the pesticide treatment gives an advantage to individuals possessing the ability to resist pesticide toxicity, and, therefore, following each pesticide treatment, the proportion of resistant individuals is increased in the population. In most cases that have been investigated, resistance is conferred by differences in a single gene between resistant and susceptible pests. How fast resistance increases depends upon how often the pesticide is used and also upon other biological factors that will be discussed below.

Now that we have discussed the way in which the frequency (or proportion) of resistant individuals increases in response to use of a pesticide, it is easy to explain a second important concept, the intensity of resistance. The degree to which any specific resistance results in reduced performance of a pesticide in the field depends upon both frequency and intensity of resistance. Obviously, resistance becomes a more serious practical threat as the frequency of resistant individuals increases in a population. Not so obvious is the fact that the intensity of resistance can be equally important in determining the practical effect of resistance on pesticide efficacy. With some resistances, the susceptibility of resistant organisms to the pesticide is only reduced by a factor of 10 to 20. Other resistances confer virtual immunity to the toxic action of a pesticide, with susceptibility reduced by a factor of 1,000 to 10,000. A resistance with a low intensity may be present in a population at a very high frequency without causing much loss of pesticide performance. On the other hand, a very intense resistance might reduce chemical efficacy when present in relatively low frequencies in a population. Now that you understand the difference between frequency and intensity of resistance, we can discuss some very different manners that the term "resistance" is commonly used.

Three Different and Common Uses of the Term "Resistance"

Based on the distinctly different way that the word "resistance" is used, it is obvious that this term means very different things to different people. To communicate effectively on the subject of resistance it is essential to distinguish between three uses of the word "resistance"—toxicological resistance, relative efficacy, and economic efficacy.

RESISTANCE. Resistance is a term derived from the perspective of the laboratory toxicologist and is defined as a heritable, statistically demonstrated decrease in sensitivity to a chemical of a pest population relative to a "normal" susceptible population. Resistance is therefore the laboratory-based comparison of differences in susceptibility between a field-collected population, which can withstand to some degree the effects of a compound, and a susceptible population (which presumably reflects pristine susceptibility conditions). Based on its scientific definition, resistance does not necessarily have any relationship to a material's performance in the field. As previously stated, a given frequency and intensity of resistance may or may not diminish the field performance of a chemical.

RELATIVE EFFICACY. Relative efficacy defines pesticide effectiveness from the perspective of the field entomologist and refers to the pest-controlling performance of a material in standardized field trials, relative to other compounds. The impact on relative efficacy of a specific frequency and intensity of resistance can be determined in standardized field trials at locations with varying frequencies of resistance (as estimated in laboratory bioassays) to the compound of interest. When resistance reduces the relative efficacy of a pesticide, it will be demonstrably less effective at locations with higher frequencies of resistance than at locations with lower frequencies of resistance, or no resistance. However, it is important to emphasize here that loss of relative efficacy of a chemical can be due to factors other than resistance. For example, microbial degradation and base hydrolysis (high pH of water) have both been known to result in reduced relative efficacy of pesticides.

ECONOMIC EFFICACY. Often, references to the presence or absence of "resistance" are based not on laboratory bioassays or field trials but on the perceptions of growers or advisors who make pest management recommendations. Economic efficacy applies to this situation and is defined as the perceived effectiveness (efficacy) of a material from the perspective of these individuals. A loss of relative efficacy occurs when users perceive that, relative to competing products, a material no longer provides economically acceptable control. A specified frequency and intensity of resistance may or may not result in a loss of economic efficacy. Loss of economic efficacy may occur without development of resistance or loss of relative efficacy; for example, it could result from the development of a product which is 10 times as active and sold at the same price per unit as competing products. Alternatively, it is not uncommon for growers to continue to use older, less expensive pesticides despite the fact that they are no longer as effective as they once were. In this case, growers perceive that the compounds provide economically acceptable efficacy despite the (often undocumented) presence of resistance and loss of relative efficacy.

Mechanisms of Resistance

It is very easy to comprehend the different general mechanisms of arthropod resistance to pesticides. The most common and powerful mechanisms in arthropods involve altered site of action of the pesticide or enhanced or altered metabolism (breakdown) of the pesticide. When the site of action is altered in resistant individuals, the pesticide does not have the intended effect on the insect's physiology. For example, insects resistant to organophosphate insecticides often possess an altered form of cholinesterase (an enzyme which is important in nerve function), which is the "target site" for these insecticides. In insects with the altered form of cholinesterase, the insecticide does not bind as readily to this target molecule. With metabolic resistance, the target site remains unchanged, but the pesticide is chemically disassembled, inactivated, or excreted before it can reach the target site. A third resistance mechanism is reduced penetration, where the absorption of the pesticide into the body is blocked or greatly limited. By itself, reduced penetration generally results in only low intensities of resistance. Combined with metabolic resistance or altered target site sensitivity, it can result in very intense resistances. Finally, a little-studied but potentially very important mechanism of pest resistance involves changes in the behavior of pests. It is very likely that some resistances involve enhanced ability of pests to avoid treated areas.

Cross- Versus Multiple-Resistance

Almost half of the reported species of insects or mites that are resistant to pesticides are resistant to two to five different classes of chemicals. Obviously, pests that are resistant to more than one of the pesticide groups registered on a crop (for example organotin and sulfite ester acaricides) pose a much greater problem to growers than do pests resistant to only one product. In discussing pests resistant to more than one pesticide, it is important to understand the distinction between cross-resistance and multiple-resistance. As previously noted, many resistances have been shown to be conferred by the action of one major gene in insects or mites.

When a resistance gene allows the pest to be resistant to more than one different pesticide, the result is cross-resistance. For example, insects possessing metabolic resistance to the organophosphate insecticide parathion are also less susceptible to some, but not all, other organophosphates. Therefore, parathion resistance confers cross-resistance to some other organophosphate insecticides. When parathion is rendered useless due to resistance, many chemically-related insecticides may also lose efficacy. The important point is that a single resistance mechanism results in resistance to more than one pesticide. With multiple resistances, however, two distinct resistance mechanisms are acquired independently through exposure to two distinct classes of pesticides. For example, some spider mites have multiple resistance to cyhexatin and dicofol. Since these resistances are acquired by two separate genetic modifications, being resistant to dicofol does not make the pest less susceptible to cyhexatin, or vice versa.

When a pest exhibits resistance to two different pesticides, it is not possible to know whether this is due to cross-resistance or to multiple-resistance without doing genetic studies of the resistant pests in the laboratory. From a practical view, cross-resistance is much more troublesome than multiple resistance, because with cross-resistance, use of one chemical can reduce effectiveness of other pesticides that the grower has never used.

RESISTANCE STABILITY AND SUSTAINABLE EFFICACY

Stable Versus Unstable Resistances

Studies have illustrated that after resistance builds up in the field, it may be stable or unstable. Stable resistances increase in populations as a pesticide is used and do not decline appreciably in populations after use of the pesticide is discontinued. Unstable resistances, in contrast, increase in frequency in response to spray applications but decrease in frequency during the period when the pesticides are not being used. For example, field studies have indicated that instability of resistance occurs in some major cropping systems around the world with the acaricides cyhexatin, propargite, and dicofol. Stability of resistance is influenced by the biological, genetic, and operational characteristics (detailed further below) of the pest, the particular resistance, and the specific agricultural system. Therefore, it is not possible to generalize from one case to another whether or not resistance will decline in a particular situation. Also, there is no way to predict, before resistance develops to a pesticide, whether it will be stable or unstable. However, in the cases where resistance is unstable, it is possible to exploit this phenomenon in formulating resistance management programs.

Minimum Treatment Intervals and Sustainable Efficacy

Because unstable resistances decline in the period between treatments with the pesticide, we can define the period of time required for the unstable resistance to decline back to the level where it was immediately prior to the previous pesticide treatment and use this as the basis for a sustainable resistance management program. The period of time that must pass for resistance frequencies to decline to pre-treatment levels (to be "reset") is defined as the minimum treatment interval. Figure 1 shows that when treatments are spaced at less than the minimum treatment interval (MTI), the resistance increases in a stair-step fashion. Alternatively, when the interval of time between treatments equals or exceeds the MTI, the resistance frequencies are able to be reset between treatments, yielding no net increase in the resistance.

Once a minimum treatment interval has been established for an unstable resistance, use of carefully selected rotations of pesticides may result in sustained efficacy of the materials in that specific cropping system. Sustainable efficacy of this sort was found in California cotton, where use of dicofol once per year has maintained efficacy of

FIGURE 1

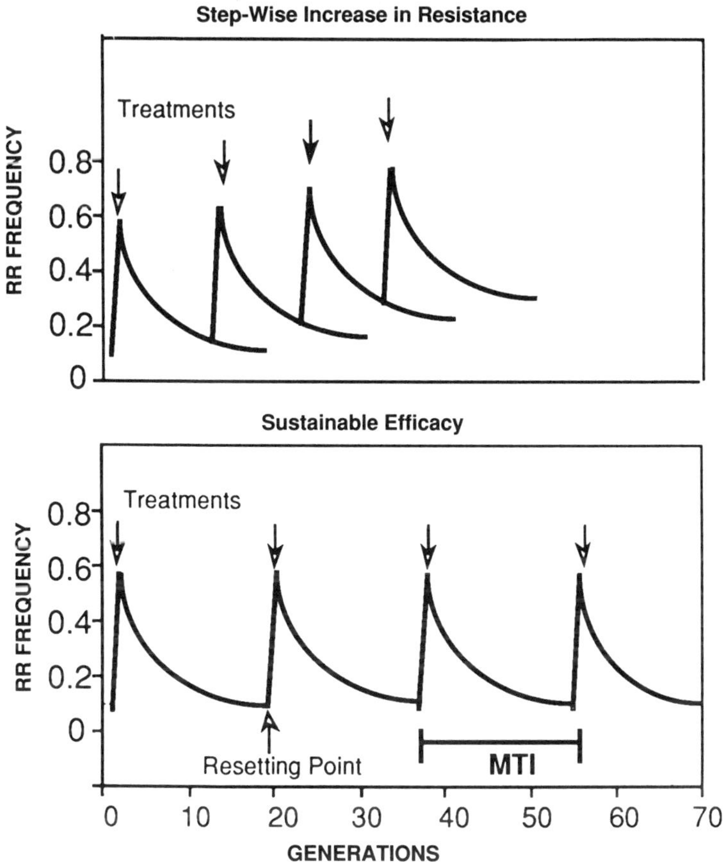

populations of *T. urticae* after over 25 years of use of this acaricide. Studies of cyhexatin resistance in apple indicated that an appropriately selected minimum treatment interval could allow sustained efficacy of organotin acaricides. Studies of mites in Brazil and Florida are developing further this concept of establishing minimum treatment intervals and appropriate rotations for acaricides. In some cases, the rate of decline of specific resistances in the field is slow. Studies of European red mites (*Panonychus ulmi*) in New York apple orchards indicated that dicofol could be used only once every three years at locations where frequencies of this resistance were high. However, as New York growers expend their few remaining registered acaricides, they are becoming more receptive to trying such long-term rotations of products.

PRINCIPLE FACTORS INFLUENCING RESISTANCE IN POPULATIONS

From an academic perspective, factors influencing resistance development can be grouped into three categories: 1) biological/ecological factors; 2) genetic factors; and 3) operational factors (*see Table 1*). Biological/ecological factors include characteristics of the pest species

TABLE 1

Factors influencing the selection of resistance to arthropods.

In-field populations
1. Genetic factors
a. Frequency of R alleles
b. Number of R alleles
c. Dominance of R alleles
d. Interactions of R alleles
e. Past selection by other chemicals
f. Interactions of R genome with fitness factors
2. Biological/ecological factors
a. Generation time
b. Offspring per generation
c. Mating system
d. Mobility/migration/isolation
e. Monophagy/polyphagy
f. Residue avoidance/refuges
3. Operational factors
a. Chemical nature of the pesticide
b. Relationship to earlier used chemicals
c. Persistence of residues/formulation used
d. Treatment threshold
e. Life stages treated of pest
f. Characteristics of the chemical application
g. Space-limited selection
h. Alternations vs. mixtures

(Adapted from Georghiou 1983)

such as growth rate and mobility, characteristics of the agricultural system such as presence of refuges for susceptible pests, suitability of hosts for pest development, the magnitude of immigration of susceptible pests, and other aspects of the integrated pest management (IPM) program such as the abundance of biological control agents. Genetic factors influencing resistance development include the number of genes conferring the resistance, the frequency and expression of resistance genes in the population, and the fitness (growth and reproduction rates) of resistant genotypes relative to susceptible genotypes. Operational factors influencing resistance are characteristics of the chemical used and the manner and frequency in which it is used, including the treatment threshold, method of application, and the use of chemical use strategies of rotations or mixtures of pesticides.

From a more practical perspective, the principle factors that influence the buildup of resistance in populations in order of importance are: 1) the degree to which you use IPM; 2) the degree to which you use your sprayer; and 3) the manner in which you select and apply pesticides. IPM practices employing, for example, routine monitoring and use of realistic pesticide treatment thresholds, use of cultural controls, and preservations or biological control agents, represent the first and most effective approach to pesticide resistance management. It is misguided to believe that resistance problems can be managed in situations where a given pesticide is used many times in a season.

CHEMICAL USE STRATEGIES FOR RESISTANCE MANAGEMENT

Three distinctly different chemical use strategies have been described by Georghiou as: 1) management by moderation; 2) management by multiple attack; and 3) management by saturation. Sustainable resistance management programs will generally incorporate multiple components of moderation (IPM tactics), multiple attack (alternation of chemicals), and in some cases, use of saturation (chemical synergists).

Management by Moderation

Management by moderation involves those chemical use practices that reduce selection on the pest population through, for example, use of low dosages of the pesticide, less frequent applications (e.g., by using higher treatment thresholds), use of chemicals with shorter environmental persistence, use of space-limited spray treatments or maintaining unsprayed areas (refugia), or spraying only specific pest stages. Many of these procedures that reduce selection for resistance are already incorporated into IPM programs and represent good pest management practices, regardless of their utility as resistance management practices. Steps taken to conserve biological controls reinforce resistance management objectives since they nearly always result in reduced pesticide selection on the pest population.

Management by Multiple Attack

Management by multiple attack involves using either mixtures or rotations of chemicals. Use of mixtures (e.g., tank mixes of two or more insecticides) for resistance management is based on the idea that organisms resistant to one of the pesticides in a tank mix will be killed by the other pesticide and that organisms resistant to both of the pesticide are rare. Though mixtures of broad-spectrum and specific fungicides have been used for years for the purpose of managing resistance (e.g., mixtures of captan and benomyl), both practical experience and theoretical models of use of mixtures lead to a strong recommendation to avoid mixtures whenever possible for control of insects and spider mites. It appears that the circumstances in which mixtures will thwart resistance development are rather rare in most agricultural systems. Importantly, when mixtures fail to prevent resistance development, the grower is often left with populations that possess high frequencies of resistance to all of the pesticides used in the mixture. This can have disastrous consequences for IPM programs.

Unless there are clear and compelling advantages to using mixtures, by far the safest strategy is to employ rotations of chemicals for control of insects and mites. When rotations are used together with components of management by moderation, maximum advantage can be obtained from those processes that allow resistance frequencies to decline in populations. Obviously, for resistance cases where instability is known to occur and where minimum treatment intervals have been established, rotations are recommended for sustaining efficacy.

Management by Saturation

Management by saturation is the strategy of using high rates of pesticide to kill resistant pests (as well as susceptible pests) or use of chemical synergists to reduce or eliminate the ability of resistant pests to metabolize a pesticide. The idea of using high rates to kill resistant pests does not work in practice under most agricultural conditions. The reason for this is that pesticide residues are deposited unevenly in most agricultural systems. Because of this, resistant pests are able to survive treatments at much higher rates than susceptible pests, even when laboratory tests show that the dosage used is sufficient to kill both resistant and susceptible types. Unfortunately, many individuals who misunderstand this strategy offer it as a logical basis for arguing against use of low rates of pesticide. The two issues are unrelated. Management by saturation does have practical application in those cases where chemical synergists can be used to "knock out" metabolic resistance mechanisms in resistant pests.

While development of practical resistance management programs is still in its infancy, cases in which various of the elements described in this section have been used can be cited, and include: 1) avoiding the use of residual pyrethroid insecticides for fly control on farms in Europe; 2) recommending alternations of acaricides in tree fruit crops in the Pacific Northwest; 3) limiting the use of pyrethroids to one mid-season application for management of resistance to *Heliothis spp.* in cotton; and 4) previously mentioned recommendations for alternations of acaricides in New York apple, California cotton, and almonds.

TWO TYPES OF RESISTANCE MANAGEMENT ACTIVITIES

The most common circumstances under which resistance management studies have been conducted are cases in which a pesticide has been used for a number of years, virtually no baseline resistance monitoring or bioassay development has ever been done, and where field failures of the pesticide are beginning to be reported. These reports of failures motivate the manufacturer to begin "resistance management" studies. This represents the *reactive case* of resistance management activities. Resistance is known or strongly suspected, usually the pesticide is a "mature" product or a new product with a mode of action common to other long-used products. The most common resistance management objectives in the reactive case begin with bioassay development and characterization of the resistance and can go on from there to multiple years of field and laboratory studies aimed at building a resistance management program. Table 2 describes the principle components of the spider mite resistance management programs under development at Cornell University.

In recent years, some chemical manufacturers have not waited until their products fail in the field to initiate resistance management efforts. They have initiated studies to develop baseline susceptibility bioassays, to look for potential problems with cross-resistance to common resistance mechanisms present in field populations and, in the absence of documented resistance, to establish "best guess" use recommendations to counter the buildup of resistance. A central part of these *proactive case* resistance management efforts is evaluation of the risk of resistance to the new chemicals by evaluating field populations from throughout the world and selecting field populations in the laboratory with the new pesticide.

PUTTING THE PIECES TOGETHER IN AN IPM PROGRAM

Now that we have outlined and defined the principle components of resistance management programs, it is important to reiterate that it is fantasy to believe that we will be able to successfully manage resistances in agricultural systems in which pesticides are used intensively. The cases cited herein in which efficacy of acaricides has been sustained for long periods of time are where the chemicals were used within the context of highly developed integrated pest management programs. There is increasing evidence to support the contention that, with good IPM practices and appropriate selection and use of pesticides, efficacy can be sustained in many systems for pesticides that are used once, or at most twice, per season. The rate at which new pesticides are being registered for agricultural use is dwindling, while currently registered pesticides are rapidly being lost to resistance. In some agricultural commodities, the day may soon arrive when materials needed to control some pest organisms are unavailable to growers because of resistance. Individuals who adopt an IPM approach (which includes a resistance management component) will have a clear

TABLE 2

Common components of a reactive resistance management program

1. Bioassay development and description of resistance
    a. compare responses of resistant and susceptible populations to a full range of pesticide doses
2. Establishment of decision-making criteria
    a. Developing diagnostic bioassay and sampling method for determining frequency of resistant organisms in field populations
    b. Choosing a critical frequency of resistance
3. Field surveys of resistance to determine geographical extent and distribution of resistant populations
4. Description of changes in frequencies of resistant populations
    a. Increase in resistance in response to selection
    b. Decrease in resistance in absence of selection
5. Description of cross-resistance relationships
6. Evaluation of chemical use strategies: treatment intervals, rotations, mixtures .
    a. Laboratory tests of resistance stability and different treatment regimes with caged populations
    b. Multi-year field trials
7. Development of key biological elements
    a. Genetics of resistance
    b. Mechanism of resistance and biochemical bioassays
    c. Fitness differences between R and S pests
    d. Augmentation of susceptible pests in the field

advantage over those who do not, and, in some cases (e.g., apples in the northeast), may be the only growers who are able to maintain susceptible pest populations.

REFERENCES

Denholm, I., A. W. Farnham, K. Odell, and R. W. Sawicki. 1983. Factors affecting resistance to insecticides in houseflies, *Musca domestica* L. (Diptera: Muscidae). I. Long-term control with bioresmethrin of flies with strong pyrethroid resistance potential. *Bull. Entomol. Res.* 73: 481-489.

Dennehy, T. J., and J. Granett. 1984. Monitoring dicofol-resistant spider mites (Acari: Tetranychidae) in California cotton. *J. Econ. Entomol.* 77: 1386-1392.

Dennehy, T. J., J. P. Nyrop, and T. E. Martinson. 1990. Characterization and exploitation of instability of spider mite resistance to acaricides. *Managing Resistance to Agrochemicals.* M. B. Green, H. M. LeBaron, and W. K. Moberg (eds). pp. 77-91.

Flexner, J. L., P. H. Westigard, and B. A. Croft. 1988. Field reversion of organotin resistance in the two-spotted spider mite (Acari: Tetranychidae) following relaxation of selection pressure. *J. Econ. Entomol.* 81: 343-347.

Georghiou, G. P. 1983. Management of resistance in arthropods. *Pest Resistance to Pesticides.* G. P. Georghiou and T. Saito (eds.). Plenum Press, New York. pp. 769-792.

Georghiou, G. P. 1986. The magnitude of the resistance problem. *Pesticide Resistance: Strategies and Tactics for Management.* National Research Council. National Academy Press, Washington, D.C. pp 14-43.

Grafton-Cardwell, E. E., J. Granett, and T. F. Leigh. 1987. Spider mite species (Acari: Tetranychidae) response to propargite: basis for an acaricide resistance management program. *J. Econ. Entomol.* 80: 579-587.

Keena, M. A.; and J. Granett. 1987. Cyhexatin and propargite resistance in populations of spider mites (Acari: Tetranychidae) from California almonds. *J. Econ. Entomol.* 80: 560-564.

Keiding, J. 1986. Prediction or resistance risk assessment. *Pesticide Resistance: Strategies and Tactics for Management.* National Academy Press, Washington, D.C. pp. 279-297.

Riley, S. L. 1989. Pyrethroid resistance in Heliothis virescens: current U.S. management program. *Pestic. Sci.* 26: 411-421.

Sawicki, R. M. 1989. Current insecticide management practices in cotton around the world: short-term successes or templates for the future? *Pestic. Sci.* 26: 401-410.

Part 3

Biology & Control of Selected Tree Fruit Pests

14

BIOLOGY, ECOLOGY, AND CONTROL OF MITES IN WASHINGTON ORCHARDS

S. C. HOYT

Washington State Universtiy

Tree Fruit Research &

Extension Center

Wenatchee, Washington

AT THE PRESENT TIME, MITES ARE NOT A MAJOR pest in most orchards in Washington State, but that has not always been the case. In the 1950's and 1960's, damage due to mite feeding was widespread and sometimes severe. It was not uncommon for growers to treat two to four or more times for mite control. Any changes in programs for insect control which would lead to the use of materials which are toxic to mite predators could again elevate mites to major pest status. Therefore, it is important to be familiar with the biology, ecology, and control of the more important species.

There are many different species of mites from several families which may occur on fruit trees, but only the common species will be discussed here.

Tetranychidae

Mites of the family Tetranychidae are generally the most important and damaging species occurring on fruit crops. The two-spotted spider mite, McDaniel spider mite, and European red mite can occur on all tree fruits grown in Washington State, but all mite species are rare on apricots. A fourth species, the brown mite, has been reported to occur, but it has seldom been found in recent times.

Eriophyidae

These very small, worm-like or wedge-shaped mites are generally the most prevalent species on most fruit crops at present. These species are difficult to distinguish, but the host plant is a clue to the most likely species. The apple rust mite occurs as a leaf vagrant on apple, but may occasionally be found on pear. The pear leaf blister mite occurs on pear and certain apple varieties and may be differentiated by its more worm-like shape and the presence of the characteristic blisters on leaves and fruit. The pear rust mite is the more common eriophyid found on pears. It feeds as a leaf vagrant but also causes extensive russeting of fruit. The cherry rust mite or plum nursery mite is found on the foliage of cherry and plum, while the peach silver mite occurs on the foliage of peach and nectarine. All of these species,, with the exception of the pear leaf blister mite, have a somewhat similar appearance and life cycle.

Phytoseiidae

Members of the family Phytoseiidae are generally predaceous, feeding on mites or small insects, but a few species also feed on pollen and fungal spores. The most common species occurring in commercial orchards of Washington is *Typhlodromus occidentalis*. It is an obligate predator, feeding on the rust mites and all stages of the tetranychids except the eggs of European red mite. Due to its predaceous habits, widespread distribution, and tolerance to many pesticides, it is an important component of IPM programs. Several other species of phytoseiids may be found in orchards not regularly treated with synthetic pesticides, the most common of these being *Neoseiulus caudiglans*.

Stigmaeidae

Zetzellia mali is predaceous on rust mites and European red mites, but I have not observed it to feed on two-spotted or McDaniel spider mites. It can be quite numerous in some orchards but generally lacks the widespread distribution of *Typhlodromus occidentalis*. This relatively slow-moving, yellow species can be important locally in IPM programs.

There are many other species of bark-inhabiting predaceous

mites which contribute to the regulation of populations of plant-feeding mites, but their physical separation from populations of the latter reduces their importance.

BIOLOGY OF THE COMMON SPECIES

Two-Spotted Spider Mite

Two-spotted spider mite overwinters as adult females (yellow to orange in color) under bark scales on the tree or in the debris on the orchard floor. Females emerge and start feeding on available green foliage of a large array of hosts in late March and April. Common orchard hosts are field bindweed, malva, and clovers, but there are many more. The females take on their characteristic straw color with two large dorsal spots and soon begin to deposit eggs on the foliage of their hosts. Moderate quantities of webbing may be found in the feeding and oviposition sites. The eggs hatch in a week or more at this time of year, depending on the temperature, and the mites go through a larval, protonymphal, deutonymphal, and adult stage. Males are somewhat spindly and diamond-shaped, while females are robust and more oval in shape. Males emerge slightly before females and tend female deutonymphs. Mating occurs as soon as the female emerges, and, in high populations, repeated matings may occur. Mated females deposit both male and female eggs, while unmated females deposit only male eggs. Each female may deposit 150 or more eggs, and there may be eight to 10 generations per year. From this, it can be seen why huge mite populations can develop so rapidly. As populations build, young females disperse either by walking to new foliage or by drifting on silken threads.

Two-spotted spider mite populations can build high numbers in the cover crop and then disperse into the tree when the cover crop matures or is mowed. However, in some instances, the mites may move directly into the trees in the spring and develop populations there.

Overwintering (diapausing) females stop ovipositing and feeding, take on the yellow to orange color, and move down the host plants to their overwintering sites. This may be triggered by shortening day length, cooler temperatures, or poor foliage quality. If populations are moderate to high, large numbers of overwintering mites may accumulate in the calyx ends of fruit, particularly apples and pears.

McDaniel Spider Mite

Generally, the life history and reproductive behavior of McDaniel spider mite is similar to that of two-spotted spider mite. There are a few important differences in the biology, however. The overwintering females are orange, and they spend the fall and winter under bark scales on the tree or in duff at the base of the tree. They have few weed hosts suitable for reproduction, so the population is more commonly confined to the tree. Therefore, the emerging females move up the tree, feed on the earliest available green foliage, change to their straw to green color, develop the four pairs of spots, and deposit eggs on the foliage. McDaniel spider mite and two-spotted spider mite can readily be distinguished in the field, primarily when actively feeding females are present. Eggs, immatures, males, and recently emerged females can be difficult to separate. McDaniel spider mite generally infests the center of the tree first, moving to the periphery and top as the season progresses and populations increase. It is more common on spur foliage and is only found on newer terminal foliage when populations are high. The species produces profuse webbing, particularly on leaves which are curled or deformed.

Where overwintered populations are high, the habit of moving directly into the tree can lead to extreme populations rather early in the season. It is not uncommon to find blossom clusters in the center of the tree completely webbed over and severely damaged. The extensive damage to foliage may lead to diapausing individuals in the first generation.

European Red Mite

The European red mite has the same life stages and reproductive behavior as the other two tetranychids, but again has many important biological differences. It can readily be separated by its color (reddish to reddish-brown or reddish-green in all stages). The overwintering stage is the egg, which is found distributed over the tree on spurs or in crevices on one- to three-year-old wood. These red, somewhat flattened eggs have a small stipe dorsally, and they hatch from about tight cluster to bloom. European red mite develops more slowly than the other tetranychids, and females lay fewer eggs. However, it has no hosts other than the fruit trees in our orchards, so it may be on the trees all season. Since the overwintering eggs are distributed over the tree, the hatching larvae move to foliage, generally on a nearby spur, and this relatively uniform distribution reduces intraspecific competition early in the season. Populations build through the spring and summer, but we have frequently observed a collapse of the population during late July or early August. We are not certain of the reason for this, but it appears that attempts to control European red mite in midsummer may prolong the infestation. When the populations collapse, few overwintering eggs are deposited, but when high numbers exist into late August or September, large numbers of overwintering eggs may occur.

Rust Mites

Since far less information exists on rust mites and their biologies generally appear to be similar, only one species, the apple rust mite, will be covered in any detail. The apple rust mite overwinters in large numbers under buds between the bud and twig and in crevices on the tree. The heaviest concentrations are usually found on one-year-old wood on terminals or watersprouts.

These mites begin to emerge when the first green tissue is visible and may form masses on this tissue at green tip. As more foliage becomes available, the mites disperse, and these large numbers are not so evident by tight cluster. Generally, the mites are wedge-shaped and have a slightly brownish tinge, although immature stages are smaller, more slender, and opaque white. The eggs are round, flattened, and difficult to see. A hand lens or microscope is essential to observe all stages. The eggs are deposited on the under surface of the leaves, and populations tend to develop on this surface until numbers are quite high.

The mites will also move to the sepals at about the tight cluster stage and may be found feeding on the fruit where the sepals join the apple at pink to bloom. On Golden Delicious, this feeding may cause calyx end russeting if populations are high. A similar, but more severe problem, occurs with the pear rust mite on pears.

There is little information on the numbers of eggs deposited by female apple rust mite or on rate of development, but there are several generations per year and very high populations can develop. Two hundred to three hundred mites per leaf occur commonly and numbers in excess of one thousand per leaf have been observed. The highest populations generally are found on the newer growth (terminals) of Red Delicious.

A population peak usually occurs in late June to mid July, with a rapid decline in numbers in mid to late July during periods of high temperatures and low humidities. Very low numbers generally persist until late August, followed by a second but lower peak in early to mid September. The mites seek suitable overwintering sites in late August and September.

ECOLOGY OF MITES

There are many factors in the physical and biological environment of mites which can affect mites and their population development. Some of these are well understood, while the information available on others is lacking or even conflicting. I will discuss some of the more important factors.

Temperature

It is well-known that high temperatures may be associated with outbreaks of tetranychid mites. This is due to a more rapid rate of egg deposit (for McDaniel spider mite, 3.8 eggs per day at 64°F versus 12.5 at 90°F), a greater total fecundity (51.5 eggs at 64°F versus 182.8 at 90°F), and more rapid developmental time (from egg to egg for McDaniel spider mite, 23 days at 64°F versus 9 days at 90°F). This results in a population doubling time of six days at 64°F versus two days at 90°F. The latter temperature was the optimum for McDaniel spider mite. Above constant 90°F, developmental time was shorter, but fewer total eggs were deposited and the oviposition rate was lower. At a constant temperature of 100°F, mortality was high. In the field, the duration of high temperatures is rarely long enough to result in extensive mortality, but it is possible to have a slight reduction in egg deposit in prolonged hot weather. This would probably be offset by more rapid development, and the net effect would be very rapid population development at high temperatures. Temperature effects on two-spotted spider mite would be quite similar to those on McDaniel spider mite.

European red mite has a lower optimum temperature, lower developmental rate, and a lower total fecundity, but the responses to temperature above and below the optimum are similar to the other spider mites.

At the other extreme of temperature, a period of chilling below 50°F is required for diapausing individuals to resume activity or for eggs to hatch. Extreme winter temperatures may result in high mortality of overwintering forms. Unusually cold weather in the spring will also cause mortality of hatching McDaniel spider mite larvae.

Typhlodromus occidentalis has its optimum for fecundity rate and developmental time at 90°F. However, total fecundity is similar from 70°F to 90°F. It is more adversely affected by temperatures above 90°F than McDaniel spider mite or two-spotted spider mite, suggesting the latter two species may be able to survive predation in periods of extremely high temperatures. However, we have not observed cases where the effects of high temperature have been responsible for decreased predation.

Humidity and Temperature

I have already mentioned the combined effects of high temperature and low humidity on apple rust mites. Very high humidity and temperature favor two-spotted spider mite over McDaniel spider mite, and very high humidity and moderate temperatures favor

European spider mite over McDaniel spider mite since McDaniel spider mite is adversely affected by high humidities.

Rainfall

Rainfall or overtree irrigation can be detrimental to McDaniel spider mite because it breaks up the webbing, creates an unfavorable micro-climate, and washes the mites off the leaves. McDaniel spider mite is rarely a problem in orchards with overtree irrigation. In contrast, European spider mite may be favored by these conditions, as rain or irrigation lower temperatures somewhat, create a favorable humidity environment, and reduce competition.

Effects of Host Plant on Mites

There are a number of ways in which the host plant can affect mites and their populations. In many cases, it may be difficult to determine which factor is the most important. The chemical constitution of the host, its nutritional status, physical features of the leaf surface, and the age or conditioning of the foliage may differ with plant species or even varieties and thereby influence the mites. Water stress alters the chemistry of the plant and may make it more favorable for mites, but if the water stress is extreme, it may retard the development of mites. These effects may vary with the species of host and mite. Increases in mite populations have been associated with nitrogen fertilization. However, conflicting results have been reported for applications of phosphorus and potassium. I believe it is safe to say that mites can develop substantial populations on any fruit tree if the tree has adequate vigor to produce a crop. The level of population in the absence of other mortality factors will be determined by the species' or varieties' tolerance to mite damage.

Effects of Mites on the Host Plant

Photosynthesis and carbon dioxide assimilation are reduced on mite-damaged leaves. This may be due to the destruction of chloroplasts and reduction in chlorophyll in the leaves. However, with certain species of mites, the disruption of the cuticle during feeding may lead to excessive cuticular transpiration, resulting in stomatal closure. It has also been suggested that damage results from an imbalance in the growth regulator complex. Regardless of the mechanism, a reduction in photosynthate is likely to first influence fruit. Subsequent effects would be on the tree's ability to withstand or recover from other stresses, and eventually roots would be starved.

Other than the direct damage to fruit of Golden Delicious by apple rust mite, pears by pear rust mite, and apples and pears by the pear leaf blister mite, the damage from all species is indirect. Extensive feeding on foliage may cause a number of effects which ultimately influence the fruit or crop load. These include but are not limited to abscission and drop of fruit or leaves, poor fruit color, reduced fruit size, improper maturation of fruit, and effects on the uptake of nutrients or plant growth regulators. There are also several reports in the literature of improved fruit color on trees where moderate mite populations occurred. High mite populations in one year can reduce bloom and set of fruit the following year. Prolonged high populations can affect tree and root growth and the general vigor of the trees.

However, there is a considerable tolerance in the trees to mite damage. The tolerance is affected by the tree species and variety, the time of year the damage occurs, the vigor of the trees, crop load, weather conditions, and other stresses on the trees. Therefore, there is no single tolerance value of mite density which fits all situations.

Effects of Agricultural Chemicals

The most obvious effect is the kill of mites by acaricides. Another major factor is the elimination or reduction of natural enemy populations, allowing the mite numbers to increase free of predation. However, with the wide array of chemicals used in orchards, there are other important but more subtle effects. Fertilizers, nutrients, and plant growth regulators alter the nutritional status of the tree, generally improving the nutriment available to the mites. Other chemicals may irritate or repel the mites, causing them to disperse from the tree or to disperse within the tree. Herbicides used on the cover crop may cause the mites to move into the trees. Several insecticides have been reported to stimulate increased egg production in mites. Excessive deposits of inert residues (much like dust) may increase the biotic potential. Finally, the use of certain chemicals may release a species from competition. Most of these effects tend to increase mite populations, but the increase may be small and seem insignificant. However, if the effect is prolonged, the resulting increase can be substantial.

Competition

Populations of a mite species become self-limiting if other regulating factors are not operating. This may be exhibited by the production of the overwintered stages, the mites dispersing from the host, a decrease in the oviposition rate, an increase in the percentage of nonviable eggs, and an increase in the mortality of immatures. This intraspecific competition does not usually regulate populations until they are well above those causing economic loss. It occurs at the lowest density with European red mite, the species most influenced by foliage condition.

Interspecific competition has not been extensively studied, particularly with the mite complex present in Washington State. Either European spider mite or two-spotted spider mite can be dominant, depending on which species infests the tree first. Webbing of either two-spotted or McDaniel spider mites will trap many European spider mites. This, plus the lower tolerance to poor foliage condition in European spider mite, generally gives the competitive advantage to two-spotted spider mite or McDaniel spider mite. Even extensive prior feeding by the apple rust mite will reduce the competitiveness of European spider mite. Some field observations suggest that McDaniel spider mite will dominate in competition with two-spotted spider mite, but the initial conditions were not known. Evidence of interspecific competition, like intraspecific, only becomes obvious at relatively high densities.

Predation

At the present time, predation by *Typhlodromus occidentalis* dominates the dynamics of mite populations in most apple and stone fruit orchards in Washington State. The presence of rust mites in moderate numbers on these crops supports the predator population. These predators prevent outbreaks of the tetranychid mites. Predators are highly effective because they distribute with the mites, are dependent on them as a food supply, they respond numerically to the mite population, and they attack more mites as the mite population increases. *Zetzellia mali, Stethorus picipes,* and a number of species of generalist insect predators prey on mites, but these species are sporadic in occurrence and do not provide the stability to the system given by *Typhlodromus occidentalis*. That stability can be disrupted by use of chemicals toxic to the predators, so it is essential to use the selective compounds in order to maintain biological control.

Other Mortality Factors

For the most part, mortality factors other than those discussed above are nonexistent or poorly understood. Several mite species are attacked by fungal or virus diseases, but to my knowledge, there are no reports of epizootics in Washington State orchards.

Mite Outbreaks and Their Causes

The potential for mite outbreaks is present because of the high reproductive potential of the mites, short life cycle, and large number of generations per year. Prolonged periods of high temperatures increase the peak population when an outbreak occurs but do not alter the potential for an outbreak. For example, using the two-day period for McDaniel spotted mite populations to double at 90°F, a population of one mite per leaf can increase to 256 per leaf in 16 days. Even at 75°F with a doubling time of 3.5 days, one mite per leaf can increase to 256 per leaf in 28 days. Of course, the mites rarely achieve their full potential for population increase. However, all that is required for a mite outbreak to occur is for the most important mortality factors to be relaxed. In Washington State tree fruits, the dominant mortality factors are predation in most situations or use of acaricides in some. Another factor which can lead to a mite outbreak is the presence of excessive field dust on foliage. Dust not only creates a more suitable physical environment for mites, it inhibits the activity of predators.

CONTROL

Reliance on chemicals as the principal means of controlling mites has several problems. Good chemical control requires relatively complete coverage of the trees. With large, older-style trees, this coverage is difficult to attain, but newer, high density plantings with smaller trees lend themselves to better coverage. However, the use of acaricides, even under the best conditions, may be self-perpetuating. You spray because you have a problem, and you have a problem because you spray, and so on.

Because of the reproductive potential of mites, nearly 100 percent kill is required or repeated applications of acaricides are essential. This places very high selective pressure on the population, and relatively rapid selection of resistant strains of mites has commonly occurred. The best approach to prevention or delaying resistance is to rely on biological control whenever and wherever it is feasible. Because we have a highly effective predator and a relatively consistent food supply for it, biological control is generally effective, cheap, and stable. Protecting predators from destruction by chemicals is the critical action.

The future of acaricide availability, like that of other chemicals, is uncertain. Based on the registration of new pesticides over the past 15 years, it may be unrealistic to expect registration of any new acaricides. Therefore, the future lies in maintaining an effective biological control program.

15

CONTROL STRATEGIES FOR LEAFMINERS AND LEAFHOPPERS REVISITED

ELIZABETH H. BEERS
Washington State University
Tree Fruit Research &
Extension Center
Wenatchee, Washington

YOU'VE HEARD QUITE A LOT ABOUT WESTERN tentiform leafminer and white apple leafhopper for the past few years from me, and despite my best efforts (in both research and extension), the problems have not improved. I've spent much time in presentations covering life cycle and biology, so I'm sure you are all very familiar with these. I'm going to skip the details here, and step back and take a look at "The Big Picture" in other words, what have we been doing, and why has it worked (or not worked, as the case may be).

I'm going to start out with a philosophical digression on what exactly a control strategy is. I think a strategy would be formulated in part by answering the following questions: 1) How much damage can be tolerated? 2) What stage of the life cycle will be targeted (timing)? 3) Will a primarily chemical or biological approach be used? 4) If a chemical approach is used, how will it be integrated with the rest of the program? 5) How will we sample for the pest? 6) Will control measures be localized or area wide? 7) How will we keep the level of pest control stable, prevent resistance from developing, or beneficials from dying off?

When we first start dealing with a pest, we tend to think in terms of "control at any cost!" (which is a pretty short-term view). Leafminer and leafhopper have been around long enough for us to start looking at the long-term view. As I review the two pests in question, I'm going to try to answer some of the strategic questions, but please keep in mind that strategies can change as we get more information. I'll also try to show some of the reasoning behind a given strategy, so you get an idea of why as well as what.

LEAFMINER

Thresholds

Leafminer has been around in most areas for five to 10 years, and we are starting to think more creatively about this pest. Let's start at the top of the list of questions with "How much damage can be tolerated?" Remember, this is an indirect pest, and not a quarantine issue, so we know we can tolerate some leaf damage. The information we have is based on one year's data in Washington State, and that was the very large and sunburned crop of 1987, so we don't have real complete information on injury thresholds. We saw some decrease in fruit size in 1987, somewhere between 1.8 and 5.5 mines per leaf. Based on this, we set a somewhat arbitrary threshold of five mines per leaf for season-long damage (cumulative mines at the end of the third generation).

The threshold for the first generation is one mine per leaf; however, we can only catch this generation before bloom because of the restriction on Vydate (it cannot be used within 30 days after bloom because of fruit thinning). At that time, we are looking at eggs, and we have found it takes about three eggs per leaf to produce one mine per leaf (because of dilution of the eggs through tree growth). The second generation threshold is two mines per leaf, and this is the critical generation both to sample and to control.

There are several reasons for this. 1) Waiting to control second generation versus first gives the parasites a chance to build up and exert control over western tentiform leafminer. 2) We have not seen a first generation population that exceeded the one mine per leaf threshold. 3) Controlling the second versus the third prevents damage during the main part of the growing season. 4) Control during June obviates the need for late season sprays that might leave detectable residues. 5) Life stages are more synchronized during second generation than during third generation.

The threshold for third generation is five mines per leaf (this is a cumulative total of all mines present at the time). So far, we have not run into a situation in which control of the fourth generation would be required, although it may be considered with late season varieties. It is likely that most of these larvae will die before reaching the pupal stage (the stage required for overwintering successfully), so late season control will not likely prevent either fruit damage or a large population the following spring.

Sampling

If you have seen a large adult population in your orchard during the prebloom, you may wish to sample for eggs. Choose 50 leaves from throughout the block from the unfolding fruit cluster leaves. Examine them with a hand lens, or better yet a microscope, and count the number of eggs, then divide by the number of leaves.

Sample the second generation during the sap-feeding stage, which should occur during late June to early July. Choose one spur leaf and one shoot leaf from 25 trees per block (or 50 leaves/block). Get a total number of mines, and divide by the total number of leaves. The same procedure can be followed for the third generation, which occurs in late July to early August. If the third generation occurs later than this (September), no control should be necessary.

Biological Control

We have a fairly effective parasitoid, a Eulophid wasp called *Pnigalio flavipes*. This wasp feeds externally on leafminer larvae, usually the tissue feeders, but less commonly, attacks sap-feeders, and the occasional pupa. It has three or more generations annually, and actually develops faster than does western tentiform leafminer. Sampling *Pnigalio* is best accomplished during the peak tissue-feeding stage. Choose 50 leaves per block, or two per tree from 25 trees. For the first generation, this sample should be taken during late May to early June. The second generation can be sampled in the same way, at the same stage of western tentiform leafminer development; this should occur during mid to late July.

If parasitism is greater than 30 to 35 percent, there is a good chance that biological control will occur, keeping western tentiform leafminer below the threshold. *Pnigalio* can cause up to 90 percent mortality of a generation of western tentiform leafminer, and if it is present in sufficient numbers, it can control western tentiform leafminer below the level at which control is necessary.

We have another parasitoid that is less common than *Pnigalio*, but can be significant in some areas or years. This is the pupa of *Sympiesis*; it is lighter in color (tan), has a long arrow-shaped abdomen, and red eyes. It is in the same family as *Pnigalio*, and is also an ectoparasite.

One tactic that has been used commonly in the past (with varying degrees of success) is importing an exotic parasite. A braconid wasp, *Pholetesor sp.*, occurs in eastern United States apple orchards, and is an important parasitoid of a closely related species of leafminer. We are considering releasing *Pholetesor* in Washington State to see if we can supplement our current cadre of parasites.

Chemical Control

Chemical control is an option with western tentiform leafminer, although we feel only a small percentage of the orchards need to be treated in any one year. In the case of this indirect pest, we like to think of biological control as the primary means of control, with chemical control only as a back-up should biological control fail. Vydate is currently the only material we are recommending for Washington State, although pyrethroids and Lannate are used in the eastern United States. However, these materials would severely disrupt our integrated mite control program, so we do not feel they are worth the trade-off.

FIGURE 1

Timing of spray applications for western tentiform leafminer in relation to life stage present and their relative susceptibility. A=time to sample mines to determine percent parasitism; B=time to sample mines per leaf to determine if threshold has been exceeded; ---indicates optimum spray timing for the second and third generations.

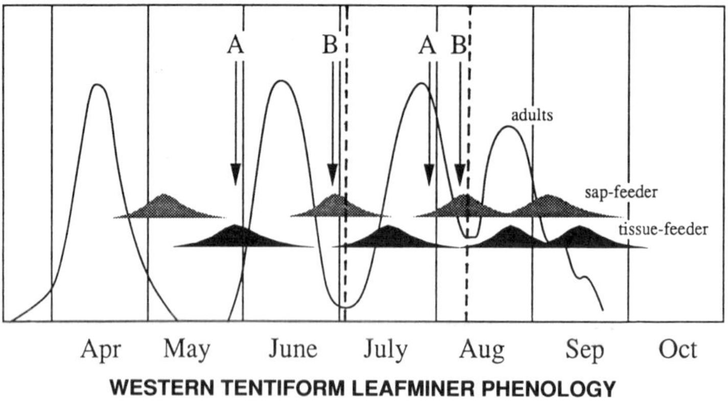

Vydate is toxic to the Western predatory mite, *Typhlodromus occidentalis*, and has also disrupted our integrated mite programs. Using Vydate at a rate of one to two pints per acre decreases the degree of disruption, but we have seen mite flare-ups in many orchards after one or two applications of Vydate. This is another reason to avoid unnecessary applications.

Vydate is toxic to all stages of western tentiform leafminer except the pupae, although it will kill only a percentage of the tissue-feeders. In order to minimize the nonsusceptible stages hit, we have recommended that applications be made during late sap-feeding stage (*see Figure 1*). This timing should be best for avoiding pupae of the previous generation, and tissue feeders of the current generation.

Vydate is toxic not only to *Typhlodromus*, it is also toxic to *Pnigalio*, along with a number of our other orchard insecticides (*see Figure 2*). Thiodan, parathion, Lorsban, and Phosphamidon also killed 100 percent of the adult females at one-half the field rate. Note that laboratory studies such as this one do not give the entire picture. The total impact of a field application will be determined by a number of interacting factors, such as length of residues, contact toxicity (the lab studies were done with topical applications), repellency, and the effect on the parasite's host. However, it is probable that the reason we see a degree of biological control in our orchards is because *Pnigalio* can survive our cover sprays of Guthion.

FIGURE 2

Toxicity of various orchard pesticides to *Pnigalio flavipes* adult females at one-half normal field rate.

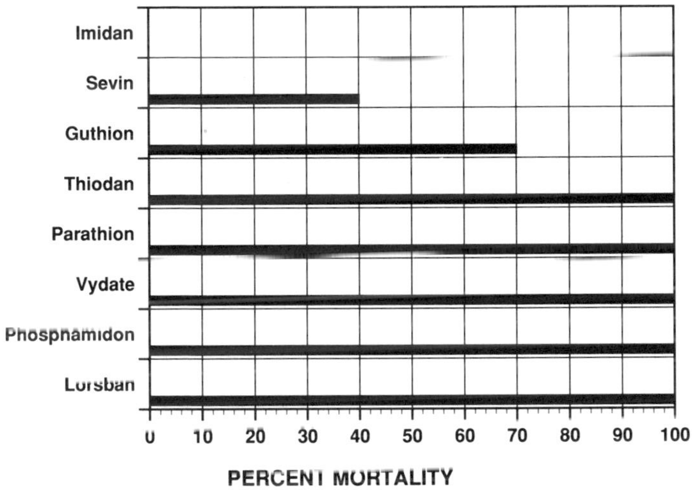

There has been some discussion of an area-wide control strategy for western tentiform leafminer. We hesitate to promote this, primarily because only a few orchards will require chemical control in any given year. Area-wide control programs could also very effectively wipe out the parasite complex over a large area. However, it is true that reinfestation is a problem, because of the mobility of the adults. It has been noted that orchards sprayed at prepink for the first generation tended not to look any different than the unsprayed neighbor by the end of the season. There is less of a chance of reinfestation occurring after application if the evaluation of the block (and control measures if necessary) are delayed until second generation.

Stability of Control

Researchers in Oregon have documented resistance of leafminer to Guthion, and this is probably also the case in Washington. We suspect that this resistance is the reason to the upswing in leafminer populations over the past 10 years. Currently, our chemical control choice consists of only one material, which makes us very vulnerable to resistance in the future. Overuse of this material is an invitation to promote resistance. However, biological control used to its maximum potential can counteract this trend. This is why we recommend we use all the tools and techniques at our disposal (especially sampling and thresholds) to promote biological control, and reduce pesticide use.

WHITE APPLE LEAFHOPPER

I'm going to switch gears now and talk about white apple leafhopper. Some of our control strategies are going to be similar, since this is also an indirect pest. Although leafhopper has been around for many years, it has become a chronic problem only within the last 10 years or so. We speculate that it was held in check by our organophosphate cover program, and only became a problem when it became more or less resistant to these materials.

Thresholds

Again, I like to start off with the threshold, because it really sets the tone for the rest of the program. I studied leafhopper damage in an orchard for two years, and found no consistent effect on fruit size, color, or quality, nor any effect on return bloom or set. In a perfectly logical world, I would end my discussion here, because if there is no damage, there is no need to sample or spray. But, one of the difficulties with testing for effects of injury by indirect pests is that the results in any one year or any one orchard may not hold true in another year or orchard. I'm relatively confident there were no measurable effects in my test orchard in either of those years, but I'm less confident that there will never be any effects in any of your orchards. So, I'm going to cover the rest of the control strategy, but hopefully, you will gain some appreciation that moderate leafhopper populations may not merit a spray application.

One of the main complaints against leafhopper is that it is a nuisance pest, especially the adults during harvest. They can be so numerous that they interfere with the pickers and impede harvesting operations. Since this is a matter of judgment, it is difficult to set a threshold for this. Each grower will have to decide how important the "picker annoyance" problem is for his or her own operation.

If picker annoyance is something you have determined you want to avoid, it would be nice to control leafhoppers before mid September, since control of adults is not very successful. I studied one orchard for several years, and found no correlation between first or second generation nymph populations and adults during harvest (a correlation would allow us to predict which orchards would require treatment to avoid excessive adult populations at harvest). I don't have a good explanation for this lack of correlation at this point, but I suspect that it is caused at least in part by movement of adults between neighboring blocks. I'll discuss this more under timing of control.

Another situation where you may want to consider control of high populations of leafhoppers is when you have several leaf-feeding pests present at the same time. If you were sustaining damage by leafhopper, mites, and leafminer simultaneously, this may cause economic damage, whereas any one pest alone does not.

Sampling

We have developed a preliminary sampling protocol for leafhopper nymphs. This method is direct visual examination of the leaves, done in the field, without removing the leaf. We recommend taking 10 leaves from 15 trees per block, counting the total number of nymphs, and dividing by the number of leaves (150). Because leafhopper populations are somewhat aggregated (clumped), many of the leaves will be zero, and a count of this type takes relatively little time. If you do plan to spray for leafhoppers, you'll probably want to assess the level of control you got, and counts before and after the spray application will give you a good idea of how successful you were.

The adults can be sampled with yellow sticky panels, which are unbaited apple maggot traps. Since this is not the target stage, this type of sample is not nearly as important as the nymph sample. You can sample overwintering eggs by partially dissecting them out the wood, but I don't really recommend this for general purposes (it is very labor-intensive). It does have one advantage, though, in that we can assess the level of parasitism at the same time we assess leafhopper populations. This may be more important in the future.

Biological Control

Leafhopper eggs are attacked by a parasitic Mymarid wasp called *Anagrus*. They have built a fairly successful program around this wasp in California for control of leafhoppers on grape, but we have not explored the full potential for biological control in Washington State. In abandoned orchards, very high rates of parasitism can occur, but in our sprayed orchards (even with a minimal 2-Guthion cover spray) we have only seen about 20 percent parasitism of the overwintering eggs.

There are several species of nymphal-adult parasitoids that occur in leafhoppers infesting cherries in The Dalles, Oregon, but we have never found them here in Washington State. At some point in the future, we may want to consider importing and releasing some of these to augment our biological control efforts.

Chemical Control

In general, for chemical control I believe we should target the nymphs, and primarily the younger nymphs (*see Figure 3*). The older nymphs and adults tend to be harder to kill, and you get less "bang for the buck" spent on insecticides. There are two chances to do this, first or second generation. It is usually easier to do a good job of cleaning up a block during the first generation, but there may be reasons for waiting until second generation (primarily reinfestation from neighboring blocks).

We have several chemical choices for control of the first generation, but all of them have some drawbacks. Phosphamidon will no longer be available; Carzol is excellent on leafhopper but toxic to predatory mites. Sevin, used at rates for thinning fruit, will kill leafhoppers, but usually the method of application (canopy sprays or skipping trees) is not conducive to good coverage. Sevin is also toxic to *Typhlodromus*. Thiodan has been consistently effective in my plots, and has relatively few undesirable side effects. However, I have heard several reports from field representatives that Thiodan is no longer effective in some orchards. I did a field study at several sites around Washington, and of the four I studied, I found one orchard where control with Thiodan was decidedly poorer than the other three (*see Figure 4*). Safer's soap is

FIGURE 3

Timing of spray applications for white apple leafhopper in relation to life stage present and their relative susceptibility. Dashed vertical lines represent optimum spray timing for first and second generations, respectively. I = first instar; II = second instar; III = third instar; IV = fourth instar; V = fifth instar.

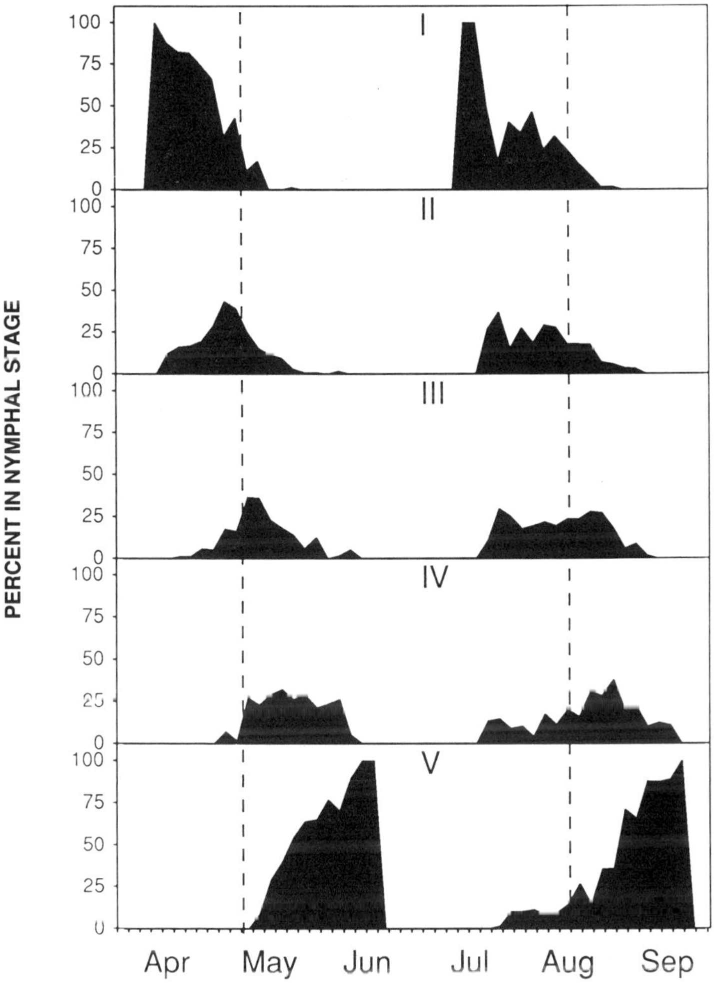

a potential new choice for leafhopper control that I am currently studying. It will not provide the same high mortality that our conventional pesticides will, but it may be sufficient to knock them back to an acceptable level. Soaps provide a way of "softening" our pesticide

FIGURE 4

Percent control with Thiodan (3 lbs/acre) at various orchard locations in Washington State.

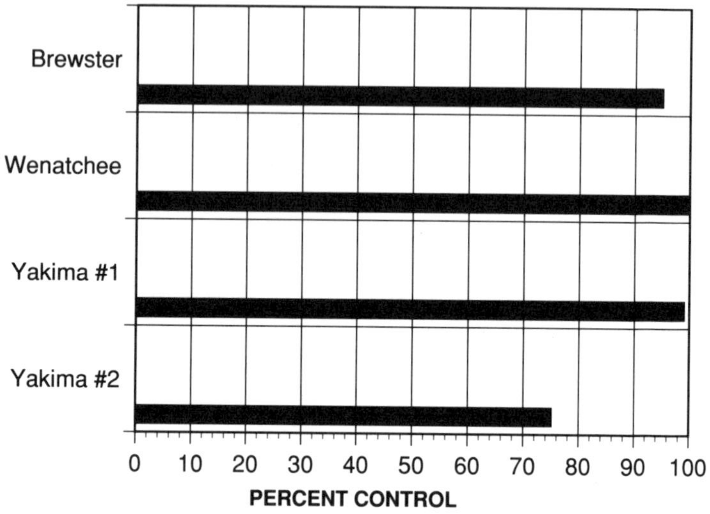

programs, and would be particularly useful as a component of a complete soft pesticide program.

Essentially, we have the same choices for second generation as for first generation, with most of the same drawbacks. Sevin can be used at this time (mid August) without fear of affecting fruit thinning, and is very effective. The other carbamate in our arsenal (Carzol) is equally effective, but both share the same high level of toxicity to *Typhlodromus (see Figure 5)*. Despite this toxicity, I have not had a mite flare-up in any of my test plots in four seasons of using these materials against leafhopper. Thiodan is still a viable choice for second generation, and predators will survive this material relatively well *(see Figure 5)*. Soap will probably not be a good choice for second generation, because of its very short residual. It kills most of what it hits, but hatch of second-generation nymphs is stretched over a long period of time, and it would take several applications to cover this period. In addition, the registrants do not recommend spraying during hot weather, and this is very likely to occur in mid to late August. Phytotoxicity is a potential problem with soaps at any time of the year, but I have not observed any in my test plots on Oregon Spur in two years.

I have done some limited work on dispersal of adult leafhoppers, in an attempt to answer the question "when is the best time to control leafhoppers?" There are several corollary questions: Does control of the first generation mean that the orchard will be clean for the rest of the season? If a neighboring block is infested, will it re-infest the one I've sprayed? If I have relatively few nymphs during August, will my adult populations be low during harvest? All these are aspects of the same question, that is, how far do adult leafhoppers move and lay eggs? From preliminary data, it appears that first generation adults will disperse readily from heavily infested blocks into clean blocks, and cause higher second generation populations. I worked with about a 30-acre block that was treated (successfully) for first generation nymphs, but still had moderate nymph populations during second generation, and fairly high adult populations in September. This was apparently because several neighboring blocks were infested.

Stability of Control

We have to be concerned about stability of control, since there is good evidence of resistance to organophosphates, and some evidence of resistance to Thiodan (an organochlorine). The carbamates are effective for now, but we can assume that leafhoppers can also develop resistance to them if they are used extensively. Exposing them to Sevin thinning sprays on an annual basis may increase the likelihood of this occurrence. Biological control is a possibility for the future, but we must change some factor (probably spray programs) in our orchards to allow more survival of parasites.

FIGURE 5

Toxicity of various orchard insecticides to *Typhlodromus occidentalis* (July and August)

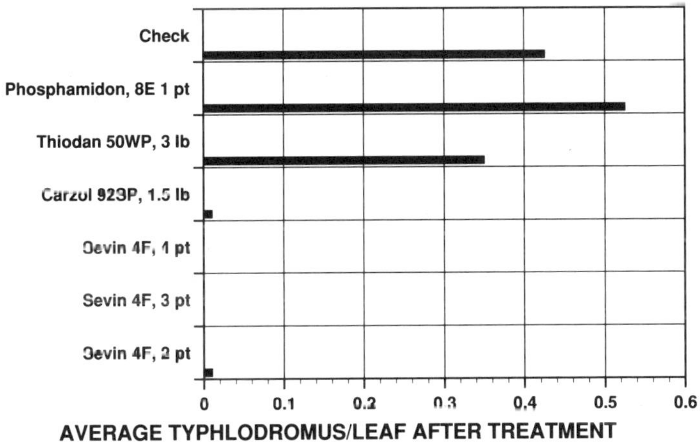

16

BIOLOGY AND MANAGEMENT OF APHIDS ON APPLE

DOUGLAS PFEIFFER
Virginia Polytechnic
Institute and State University
Blacksburg, Virginia

MOST OF THE APHIDS THAT ARE PESTS OF FRUITS are holocyclic, meaning that they spend the winter as an egg (the result of a sexual union) on the host plant. This over-wintering host is the primary host. The eggs of the various species are indistinguishable. In spring or summer, the aphids reproduce asexually, giving live birth to daughter aphids, often on a different host plant species, the secondary host.

Aphids and some related Homoptera feed on phloem sap, generally from growing shoot tips where nutritive quality is highest. A few, such as woolly apple aphid and grape phylloxera, feed on galls with less organized tissue cellular structure, and their stylets cannot reach the phloem. These species feed on cellular juices after cells are disrupted.

The impact of individual aphids on their host plants is generally very slight. Their importance comes from the numbers of aphids involved. This is important for Integrated Pest Management, since population densities are often below economically damaging levels. Some species, however, carry an importance out of proportion to their numbers, because of host plant effects beyond simple removal of assimilate, including toxic saliva, production of galls or other distortions of growth, or transmission of viruses.

WHAT APHIDS ARE ON APPLE?

Many aphid species have been recorded on apple from various parts of the world (*see Table 1*), causing various degrees of leaf abnormality. Because of their sheer numbers, they have been assumed to cause important damage. There have been recent changes in the species composition of common apple aphids, so it is worth reconsidering the composition of the aphid complex on apple. Our later discussion will concentrate on rosy apple aphid, apple aphid, spirea aphid, and apple grain aphid.

TABLE 1

List of aphid species reported from apple.

Allocotaphis quaestionis (Börner)
Aphis craccivora Koch
Aphis gossypii Glover
Aphis pomi DeGeer
Aphis spiraecola Patch
Dysaphis anthrisci Börner
Dysaphis brancoi (Börner)
Dysaphis chaerophylli (Börner)
Dysaphis devecta (Walker)
Dysaphis plantaginea (Passerini)
Dysaphis radicola (Mordvilko)
Eriosoma lanigerum (Hausmann)
Macrosiphum euphorbiae (Thomas)
Myzus persicae (Sulzer)
Nearctaphis bakeri (Cowen)
Ovatus crataegarius (Walker)
Ovatus malisuctus (Matsumura)
Rhopalosiphum insertum (Walker)
Toxoptera aurantii (Boyer de Fonscolombe)
Watabura nishiyae Matsumura

(*Blackman & Eastop 1984*)

Rosy Apple Aphid, Dysaphis plantaginea (old name Aphis sorbi)

APPEARANCE. The appearance of this species varies; it is typically thought of as pink, but it is often purplish brown. The body is covered with fine white powdery material. The cornicles and antennae are black.

GENERAL. Rosy apple aphid was apparently introduced into North America in the mid 1800's, but didn't reach pest status until late in that century. Its spread paralleled that of the main summer host, narrowleaf plantain, which was introduced early, in the late 1600's, but spread slowly.

Most growers are familiar with the curled leaves caused by this species. Possible benefits of this curling to the aphid have sometimes been discussed. One benefit may be protection from natural enemies; this is probably not very important. Ants are often seen tending colonies of rosy apple aphids. Parasites and predators likewise find little obstruction from curled leaves and are known to attack rosy apple aphid. The aphids probably get some protection from rain, which can sometimes wash aphids off their host plant. A problem related to this is that the aphids are also protected from contact with insecticidal sprays, leading to difficulties in control after aphid colonies are established.

There is a further benefit from aphid-affected leaves for aphids. Curled leaves are a better food source for the aphids. Forrest put *Aphis fabae* Scopoli on apple seedlings with and without *Dysaphis devecta*, a species related to our rosy apple aphid and causing a similar leaf curl. Without *D. devecta, A. fabae* got smaller and smaller with each succeeding generation. In the presence of *D. devecta, A. fabae* maintained its body size.

Rosy apple aphid also causes leaves to curl. Twigs curl toward feeding aphids; curling may arise from multiple colonies or perhaps from the twisting pattern of phloem tissue in the twigs. This is an important source of damage in young, nonbearing blocks.

Fruit may also be stunted and misshapen, important in bearing blocks. This direct form of injury is not important in other aphid species on apple.

LIFE HISTORY. Rosy apple aphid overwinters in the egg stage. Eggs hatch around the green-tip to half-inch-green stages of bud development, about a week later than apple grain aphid, and about the same time as apple aphid. As soon as they hatch, nymphs enter the unfolding buds. Research in England has shown that leaves will curl in response to a single nymph feeding for only 24 hours

Stem mothers mature after about two weeks. Rosy apple aphids are prolific, averaging 133 nymphs per female (5.4/day/female);

second generation aphids average 119 nymphs per female. The fifth instar is the vivipara. Populations are usually greatest in the interior and upper parts of the canopy.

Rosy apple aphid passes through only three to four generations on apple. The majority of third generation aphids are alates (having wings) and disperse to summer hosts. Alates are reluctant to feed on plantain, are attracted to light, and display a need to undergo a dispersal flight. These aphids are more active, dispersing from the plant more readily. Subsequently, rosy apple aphid breeds continuously on plantain. After progeny are produced on this plant, they look like a different aphid species, now appearing pale yellow.

About six generations occur on the summer host. No noticeable symptoms appear on plantain, and the aphids are at low densities. Hot dry weather is detrimental to rosy apple aphid populations. This is a poorly known part of the life cycle. The importance of this aphid species varies considerably from year to year and from orchard to orchard. Perhaps this unexplained variability arises in part from natural sources of mortality during the approximately five months when the aphids reside on plantain.

In late autumn, winged migrants are produced which fly to apple from late September throughout October. Males appear later, and are active into November. The fall migrants look like spring alates, but are darker. These migrants give rise to oviparae, which are sexual, egg-laying females. Oviparae are yellowish with a pale green tinge. Oviparae move to smaller twigs, where they mate with males and lay eggs around bases of buds and other protected areas. An average of four to six eggs are laid per female.

Apple Aphid, Aphis pomi

APPEARANCE. This aphid is green, sometimes called yellowish green. The comicles, tips of antennae, and cauda are black.

GENERAL. Apple aphid is the only one of the three species most commonly cited on apple to spend the entire season on the apple tree. This is the familiar aphid, often called simply "green aphid" by growers, that is most frequently found on apple trees in spring and summer. Apple aphid was described by DeGeer in 1773, in Sweden, where it was reported to often kill trees. Apple aphid was imported from Europe shortly before 1850. Asa Fitch collected it from the United States in 1851; it was found in Washington State in 1883. Good descriptions of the species were given by Baker and Turner (1916) and Matheson (1919).

LIFE HISTORY. Eggs are laid on small twigs, especially watersprouts. Baker and Turner said that the twigs could be almost black with eggs, but levels this high are rarely seen. Many eggs are apparently infertile; many workers have noted the variable and sometimes high rate of failure to hatch, saying that it was unknown if the eggs were infertile. Baker and Turner said that it is fairly easy to determine fertility, since the infertile eggs are orange instead of yellowish when first laid. This will rarely be seen however, since the eggs rapidly turn green and later black, usually within a day. Eggs develop for about five days until they reach a resting stage at which they pass the winter. In Virginia, they resume development in March. While in the resting stage, if there is a substantial warming period, many embryos die. Eggs hatch around green-tip.

The stem mother is a wingless, viviparous female. Stem mothers wander for about a day without feeding, settle, and rarely move thereafter. After averaging 10 to 11 days as a nymph (about 20 days in cooler temperatures), they reach adulthood and begin reproducing about the next day. Adults often appear around bloom time. A stem mother can produce about 40 offspring over 10 days. Soon after becoming adults (often within 24 hours), viviparae begin producing nymphs. These nymphs can develop into alates or wingless individuals, but alates are more common than in the other aphid species on apple (Matheson reported 75 percent alates). It is possible to go all the way from stem mother to sexuals without producing winged viviparae.

There are seven to 17 generations, depending on whether first or last young of each generation are considered. Alates disperse, creating colonies elsewhere in the orchard. Wingless individuals are quiet; alates are more active, feeding for shorter periods. Wingless aphids produce more offspring than alates (*see Table* 2); this is true in other aphids as well. The reproductive rate of apple aphid varies considerably during the season (*see Table* 2). In later generations, fewer alates are produced. Apple aphid populations climb in May and June, then aphid days stabilize as populations decline for one to

TABLE 2

Variation in reproduction by apple aphid.

Period	Apterous	Alates
before early July	55	50
early July-early September	31	25
after early September	24	few

(Baker & Turner 1916)

three weeks after the population peak.

Less leaf curl or fruit effects are caused by apple aphid than by rosy apple aphid. This species causes little leaf curl in lab, although this is sometimes reported in the field. It is common to have one or two rosy apple aphids in an apple aphid colony; this may be account for such field reports.

During approximately the 14th generation, males and sexual females (oviparae) are produced. Oviparae and males are wingless, smaller than other forms. Males mate throughout their activity period, but many females do not become fertilized because of a relative scarcity of males (males comprise a minority of the population and don't live as long as oviparae). During one season, oviparae began ovipositing on October 4. Oviposition can continue through the first week of December.

HOST PLANT EFFECTS. Hamilton studied the effects of aphids identified as *A. pomi* on Red Delicious and Golden Delicious trees. A commonly used threshold of 50 percent infested terminals correlated with 15.7 aphids per shoot. This study related aphid days to several plant parameters: soluble solids, shoot growth, fruit weight, and fruit numbers. The two varieties were examined over three years. On Red Delicious, aphid numbers never were associated with a reduction in any plant parameter. On Golden Delicious, there were significant effects in only one of the three years. Soluble solids were increased slightly, increasing fruit sweetness (not enough to be important for practical purposes). For every increase of 1,000 aphid days, there was a 1.5 inch reduction in shoot length.

Phosphamidon had its greatest effect at reducing aphid-days when applied between 450 to 650 degree-days starting at green tip, using a developmental threshold of 42°F. When applied before this time, aphids were likely to rise again later; when applied later, more aphid-days were likely to have accrued beforehand. Populations reaching the threshold of 15.7 aphids per shoot did not always coincide with the most effective time to spray phosphamidon.

BIOLOGICAL CONTROL. Biological control is more effective against apple aphid than rosy apple aphid. There are variety of predators. Adult and larval lady beetles, larval lacewings, and syrphid flies each have a role in reducing populations. Syrphid eggs and larvae may be seen in aphid colonies early in the season. A predatory midge, *Aphidoletes aphidimyza*, has been important in some areas. There are at least six generations annually, but the midges are not abundant early in the season. Some resistance to Guthion has developed, but activity is still greater under reduced spray programs.

Several parasites attack apple aphid, but many are unable to complete their development. Although they kill the aphid, the parasites die in the last larval or pupal stage. This hinders their effectiveness in control. They cannot respond numerically to the aphids, and so the proper surrounding habitat is needed to provide a source of immigrating parasites. Further work in this area is needed.

Ground cover factors can affect tree vigor and therefore aphid numbers, and can also affect the density of certain predators. Unfortunately, low vigor treatments with fewer aphids also had less terminal growth. Further work is needed on the role of ground cover on tree vigor, fruit quality, and the arthropod community in mature orchards.

Apple Grain Aphid, Rhopalosiphum fitchii

APPEARANCE. This aphid is bright yellowish-green with a darker green band running the length of the abdomen.

GENERAL. This aphid is often the most abundant aphid early in the season, arousing growers' concern by its dense populations clustering on tips of twigs. There may be some beneficial effect of apple grain aphids by helping to build up predatory populations before the increase in other aphids.

LIFE HISTORY. Eggs are not killed by warm periods in the winter as are those of apple aphids. These are the earliest aphids to hatch, about a week or 10 days before rosy apple aphid and apple aphid. It can be distinguished from apple aphid by the dark green stripe down the middle of the back. The first two generations feed only on apple leaves. Most leave the primary host as second generation adults; a few third generation individuals may still be found on apple. Summer hosts include many grasses and grains. Fall migrants return to apple trees, often outnumbering the other aphid species. Males arrive about a week later. Many apple grain aphids die with falling leaves.

Spirea Aphid, Aphis spiraecola, as a Major Component

In the mid 1980's, it became apparent that spirea aphid was a common aphid in apple trees in Virginia. We determined that 87 percent of "green aphid" colonies in May and June were in fact composed of spirea aphid. For years, this aphid has been a pest of citrus. It was recorded long ago on apple. Edith Patch described this aphid as a separate species. She later changed her mind and considered it a race of apple aphid, since both occurred on apple and were similar morphologically. But she underestimated the importance of some differences, and they are now considered separate species.

For years, spirea aphid was assumed to overwinter on spirea and disperse to other hosts in the spring. In 1979, spirea aphid was found overwintering on citrus. We sampled oviparae in the late fall; 73 percent were spirea aphid. We also found spirea aphid in the first spring generation; this was the first report of spirea aphid using apple as a primary host. Along with the host shift to apple, there has been a shift in egg hatch time to correspond with host plant phenology. The practical meaning of this is that spirea aphid will be on the apple tree earlier in the season. The shift to apple seems to be worldwide; spirea aphid now dominates in Israel and China. Specimens I examined from Washington State and Massachusetts included spirea aphid *(see Table 3)*. This species has made large host shifts before, including its movement into the tropics in the first half of this century.

TABLE 3

Percentage of spirea aphid in green aphid colonies in Washington State and Massachusetts.

	Spirea aphid	Apple aphid
Washington		
Smith Tract	7	2
Columbia View	6	3
TFRC	10	1
Massachusetts		
Belchertown	8	2
Roberts	8	1
Quincy	2	8

TABLE 4

Characters separating spirea aphid from apple aphid.

Viviparae	Spirea aphid	Apple aphid
Caudal hairs	12 or less	14 or more
Tubercles on abdominal segments 2-4	no	yes
Sexual generation		
Hind tibiae of oviparae	swollen	slender
Males wings	present	absent

Spirea aphid is distinguished from apple aphid by several characters, all of which are impossible to use in the field. These characters are summarized in Table 4. They are best seen in slide-mounted specimens, but can also be discerned in alcohol under a dissecting microscope.

SPIREA APHID VERSUS APPLE APHID: SO WHAT?

Effects on Apple Leaf Function

In 1986 and 1987, spirea aphids and apple aphids were placed on young apple trees in a greenhouse in Blacksburg, Virginia. Spirea aphid suppressed photosynthesis. When spirea aphid was compared with apple aphid, there was little difference in photosynthesis between the treatments.

Effects on Apple Tree Growth

We examined the effect of spirea aphid and varying nitrogen fertilization levels on tree growth. As urea application rate to trees increased, so did the number of aphids per shoot. The date of the aphids' population peak was earlier on trees with higher nitrogen.

The interaction between spirea aphid and nitrogen was examined (*see Table 5*). Feeding by spirea aphid reduces percent foliar nitrogen, while fertilization increases foliar nitrogen. Trees that are infested with spirea aphid but are fertilized may contain more foliar nitrogen than noninfested trees that receive no fertilization. Thus, nitrogen fertilization can compensate for the effects of aphid feeding.

When tree growth was analyzed according to aphid infestation status, it was found that spirea aphid feeding reduced both fresh and

TABLE 5

Effect of spirea aphid infestation and urea application on percent foliar nitrogen.[1]

	g urea per tree			
Treatment	0.0	0.5	1.0	2.0
Control	1.51c	1.55c	2.21b	2.71a
	*	*	*	*
Spirea aphid	1.38d	1.42c	1.89b	2.12a

[1]*Means followed by the same letter in a row are not significantly different. Means in a column separated by an asterisk are significantly different.*
(*Kaakeh 1989*)

dry weight at harvest, as well as net gain. When the trees were analyzed according to fertilization rate, the growth parameters increased with increasing nitrogen level. Fertilization helped trees compensate for aphid feeding, as in the case of percent foliar nitrogen.

The growth of trees infested with apple aphid was compared with trees infested with spirea aphid. There were no significant differences in dry weights of shoots, trunk, rootstock, roots, or leaves. It was therefore concluded that the two aphid species affect the tree in a similar manner, and so economic injury levels should not differ.

Response to Insecticides

Using a laboratory slide dip method, Hogmire found that apple aphid and spirea aphid differed in susceptibility to some common orchard insecticides. In the field, there were no differences between species in susceptibility to some compounds, i.e., Lannate, Lorsban, and Thiodan. However, spirea aphid was better able to tolerate Guthion and Asana. Guthion is easily the most frequently applied insecticide in apple orchards in the eastern region. This could be part of the reason for the dominance of spirea aphids in our orchards.

Potential Other Differences?

Differences in insect-plant interactions and susceptibility to insecticides between apple aphid and spirea aphid have been examined. There are other potential areas of interspecific differences. The two species have been shown to respond differently to high temperatures, with differing rates of population increase. Possible differences in effective natural enemies are unexamined to date. Spirea aphid is a dead-end host for several parasitic wasps, as is apple aphid, including some of the same wasp species.

ROSY APPLE APHID

Effects on Apple Leaf Function

As part of a project on insect-plant interactions between rosy apple aphid and apple, effects on leaf function were studied. It was difficult to relate photosynthesis to aphid numbers because of the inability to count aphids within curled leaves. So, rather, aphid feeding was expressed in terms of degree of leaf curling. Two levels of curling (180° and 360°) were examined, both of which had less than half the photosynthetic activity of undamaged leaves. Damaged leaves also contained significantly less chlorophyll than control leaves.

Several species of aphid on pecan have been shown to cause the accumulation of callose at phloem sieve plates. This could impede the translocation of phloem sap. Varn sectioned and stained leaves fed upon by rosy apple aphid and found such deposits in these leaves as well (more than five times as many as in control leaves).

Effects on Apple Tree Growth

When examining aphid effects on tree growth in this experiment, we attempted to simulate natural population levels and periods of infestation of rosy apple aphid, comparing it with spirea aphid. Equivalent densities of the two species were not compared. We used potted trees with pots sunk in ground to maintain normal root temperatures.

When tree growth was analyzed at the whole tree level, it was seen that spirea aphid reduced net gain in fresh weight, but not nearly as much as did rosy apple aphid. When dry weight of total woody tissue was examined, spirea aphid was not significantly different from the control, while rosy apple aphid had a significant effect.

Trees were divided into scion (both 1- and 2-year portions) and root (both fibrous and total). Samples were collected at harvest time. Rosy apple aphid reduced both one- and two-year scion wood (spirea aphid was intermediate). Spirea aphid had no effect on root growth, but rosy apple aphid reduced both fibrous and total root tissue.

Trees harvested at the 10-leaf stage in the year following infestation showed similar patterns. This stage was chosen because it was thought that the trees would have used most of their stored reserves before beginning to assimilate nutrients from the next season's leaves. The lack of effect by spirea aphid (in contrast with the study by Kaakeh) may be attributed to differing population densities or infestation periods.

Chemical Control

CHEMICAL. Hull and Starner compared various insecticides at different timings in relation to bloom and rosy apple aphid egg hatch. The eggs hatched from silver tip to half-inch-green. Lorsban or Supracide applied during hatch each gave season-long control. The addition of oil did not increase control, and oil alone was insufficient. Applications after half-inch-green, especially with contact insecticides, resulted in less effective control.

HOST PLANT RESISTANCE. Apple varieties resistant to rosy apple aphid are not yet available, but Alston and Briggs reported lines of *Malus robusta* with a hypersensitive response. Sexual aphids return to apple trees in the fall and lay eggs. In the spring, these eggs hatch but the nymphal stem mothers evoke the hypersensitive response and fail to establish breeding colonies.

SEASONAL APHID MANAGEMENT PROGRAM

Currently Available Tactics

ROSY APPLE APHID. A delayed dormant application of Supracide or Lorsban at one-quarter to one-half-inch green will give effective control of rosy apple aphid and the "green aphid" complex. Oil is not necessary for aphid control, but mites are not controlled by these products. Later, if necessary, Cygon or another systemic may be used for aphid populations. But care should be taken not to spray too late for rosy apple aphid; a late spray will not kill many aphids, they are about to leave the tree anyway, and the damage is already done.

When rosy apple aphid is vulnerable to sprays, population density is so low that it is very difficult to scout effectively. It is generally worthwhile to apply the delayed dormant spray for this species, regardless of perceived pest status. One season's unrestrained rosy apple aphid population on young trees can cost $14 to $20 per tree by the time the tree reaches maximum yield (aphid-damaged limbs must generally be pruned off, setting back the development of the tree). On bearing trees, direct fruit damage can be important. The cost of the spray application is less than $0.005 per tree.

"GREEN APHIDS" (SPIREA AND APPLE APHIDS). The delayed dormant application described for rosy apple aphid will also be effective for *Aphis spp*. It is probably not worth controlling summer populations of the "green aphid" complex on bearing Red Delicious trees because of the low correlation between aphid populations and fruit yield or quality. More effective biological control for aphids as well as other pests may be obtained by withholding sprays.

On mature Golden Delicious trees, use an action threshold of 50 percent infested shoots, tempered by number of shoots with aphid natural enemies. If 20 percent of the shoots contain predators of aphids (syrphids, midges, lacewings, lady beetles, etc.), do not spray, or wait one week and reevaluate. Choose insecticides which are the least toxic to beneficial species.

Aphid populations can reduce growth of young apple trees, so aphids in such blocks should not be ignored. However, if trees are kept in a vigorous condition by proper fertilization, effects of aphid feeding can be minimized.

Future Tactics

ROSY APPLE APHID. A complete understanding of the summer phase of the life cycle of this species should aid in the overall management. Perhaps we can minimize populations at this point in the life cycle, before they return to the trees. Breeding of the hyper-sensitive host response into commercial varieties would give a permanent control without the use of insecticides.

"GREEN APHIDS" (SPIREA AND APPLE APHIDS). Information is needed on interspecific differences in natural enemies. The shift in species composition on "green aphids" may seem to be minor from the perspective of natural enemies, but similar changes in the species composition of grape leafhoppers in California have recently altered the status of biological control.

Management of both aphid groups will doubtless be affected by the advent of new aphicides. Mating disruption, currently under development for a variety of pests, will be useless for summer populations because of the lack of males. But this technique is a theoretical possibility for fall populations. No data are available for aphids; the sex pheromones of these insects are poorly known.

The alarm pheromones have received some attention for applications in control. By increasing aphid activity, the effectiveness of contact insecticides may be enhanced. Biological control may be improved by either habituation to constant levels of alarm pheromone or inhibition of response to the pheromone.

REFERENCES

Adams, R. G. and R. J. Prokopy. 1980. *Aphidoletes aphidimyza*: An effective predator of the apple aphid in Massachusetts. *Protect. Ecol.* 2: 27-39.

Alston, F. H. and J. B. Briggs. 1970. Inheritance of hypersensitivity to rosy apple aphid *Dysaphis plantaginea* in apple. *Can. J. Genet. Cytol.* 12: 257-258.

Baker, A. C. and W. F. Turner. 1916. Morphology and biology of the green apple aphis. *J. Agric. Res.* 5: 955-993.

Blackman, R. L. and V. F. Eastop. 1984. *Aphids on the World's Crops: An Identification Guide.* Wiley, New York.

Brown, M. W., H. W. Hogmire, and J. J. Schmitt. 1988. Simulation of the interaction between the apple aphid and spirea aphid. Proc. Cumberland-Shenandoah Fruit Workers' Conf., Harpers Ferry, West Virginia.

Carroll, D. P. and S. C. Hoyt. 1986. Hosts and habitats of parasitoids (Hymenoptera: Aphididae) implicated in biological control of apple aphid (Homoptera: Aphididae). *Environ. Entomol.* 15: 1171-1178.

Davis, J. J. 1916. Aphidoletes meridionalis, an important dipterous enemy of aphids. *J. Agric. Res.* 6: 883-888.

Forrest, J. M. S. 1971. The growth of *Aphis fabae* as an indicator of the nutritional advantage of galling to the apple aphid *Dysaphis devecta*. *Entomol. Exp. Applic.* 14: 477483.

Forrest, J. M. S., and A. F. G. Dixon. 1975. The induction of leaf-roll galls by the apple aphids *Dysaphis devecta* and *D. plantaginea*. *Ann. Appl. Biol.* 81. 281-288.

Gibson, R. W. and A. D. Rice. 1989. Modifying aphid behaviour. p. 209-224. A. K. Minks and P. Harrewijn (eds.) Aphids; Their Biology, Natural Enemies and Control. *World Crop Pests* 2C. Elsevier, New York. 312 pp.

Haley, S., and E. J. Hogue. 1990. Ground cover influence on apple aphid, *Aphis pomi* (Homoptera: Aphididae), and its predators in a young apple orchard. *Crop Prot.* 9: 225230.

Hamilton, G. C., F. C. Swift, and R. Marini. 1986. Effect of *Aphis pomi* (Homoptera: Aphididae) density on apples. *J. Econ. Entomol.* 79: 471-478.

Hogmire, H. W., M. W. Brown, and V. L. Crim. 1990a. Toxicity of slide dip application of five insecticides to apple aphid and spirea aphid (Homoptera: Aphididae). *J. Entomol. Sci.* 25: 10-15.

Hogmire, H. W., T. Winfield, M. W. Brown, and J. J. Schmitt. 1990b. Toxicity of insecticides to two aphid species on apple, 1990. Proc. Cumberland-Shenandoah Fruit Workers' Conf., Kearneysville, West Virginia.

Hull, L. A. and V. R. Starner. 1983. Effectiveness of insecticide applications timed to correspond with the development of rosy apple aphid (Homoptera: Aphididae) on apple. *J. Econ. Entomol.* 76: 594-598.

Kaakeh, W. 1989. The effect of spirea aphid (Homoptera: Aphididae) feeding and nitrogen fertilization on the growth of young apple trees, with comparisons to apple aphid. Ph.D. dissertation, Va. Polytech. Inst. St. Univ., Blacksburg, Virginia. 184 p.

Komazaki, S. 1983. Overwintering of the spirea aphid, *Aphis citricola* Van der Goot (Homoptera: Aphididae) on citrus and spirea plant. *Appl. Entomol. Zool.* 18: 301-307.

Komazaki, S. 1990. Variation in the hatch timing of the overwintering egg among populations of *Aphis spiraecola* Patch (Homoptera: Aphididae) collected from different host plants and localities in Japan. *Appl. Entomol. Zool.* 25: 27-34.

Matheson, R. 1919. A study of the plant lice injuring the foliage and fruit of the apple. N. Y. Agric. Exp. Stn. (Cornell) Mem. 24: 683-762.

Newton, C. and A. F. G. Dixon. 1990. Embryonic growth rate and birth weight of the offspring of apterous and alate aphids: a cost of dispersal. *Entomol. Exp. Appl.* 55: 223229.

Patch, E. M. 1914. Maine aphids of the rose family. Maine Agric. Exp. Stn. Bull. 233.

Patch, E. M. 1923. The summer food plants of the green apple aphid. Maine Agric. Exp. Stn. Bull. 313.

Pfeiffer, D. G., M. W. Brown and M. W. Varn. 1989a. Incidence of spirea aphid (Homoptera: Aphididae) in apple orchards in Virginia, West Virginia, and Maryland. *J. Entomol. Sci.* 24:145-149.

Pfeiffer, D. G., J. F. Derr and R. P. Marini. 1989b. Agricultural chemical use by Virginia apple growers. Proc. Cumberland-Shenandoah Fruit Workers' Conf., Harpers Ferry, West Virginia.

Starner, V. R. and L. A. Hull. 1982. Distribution of rosy apple aphids (Homoptera: Aphididae) within the canopy of apple trees. *Environ. Entomol.* 11: 964-967.

Tremblay, E. and S. Barbagallo. 1982. *Lysiphlebus testaceipes* (Cr.), a special case of ecesis in Italy. R. Cavalloro (ed.). Aphid Antagonists. Proc. of a meeting of the EC Experts' Group, Portici, Italy.

Varn, M. W. 1987. The effects of aphids on the growth of young apple trees. Ph.D. dissertation, Va. Polytech. Inst. St. Univ., Blacksburg, Virginia. 91 p.

Varn, M. W and D. G. Pfeiffer. 1989. Effect of rosy apple aphid and spirea aphid (Homoptera: Aphididae) on dry matter accumulation and carbo hydrate concentration in young apple trees. *J. Econ. Entomol* 82: 565-569.

Wolcott, G. N. 1954. Dispersion to the tropics of the spirea aphid. *J. Econ. Entomol.* 47: 568-571.

17

LEAFROLLER PESTS OF FRUIT CROPS IN WASHINGTON STATE

JAY F. BRUNNER
Washington State University
Tree Fruit Research &
Extension Center
Wenatchee, Washington

SEVERAL IMPORTANT PESTS OF TREE FRUIT CROPS belong to the family *Tortricidae* which are commonly referred to as leafrollers. Their common name comes from the behavior of their larvae which roll or tie together leaf surfaces when constructing feeding sites or shelters. There are four species of leafroller reported as pests of fruit crops in the western United States and Canada. These are the fruit tree leafroller (*Archips argyrospilus*), European leafroller (*Archips rosanus*), pandemis leafroller (*Pandemis pyrusana*), and the obliquebanded leafroller (*Choristoneura rosaceana*). Other species of leafroller are known as pests of fruit crops in the eastern United States and Canada.

In Washington State, leafrollers were reported as pests of fruit as early as 1921. In the late 1940's and early 1950's, F. J. Newcomer and Carlson reported pandemis leafroller feeding on cherry, apricot, and apple. In about 1970, pandemis leafroller was reported causing damage on apple and cherry. During the past 20 years, pandemis leafroller has been observed with increased regularity in cherry and apple orchards of the Yakima Valley and northcentral Washington State. In 1985, recommendations for use of chlorpyriphos (Lorsban®) in delayed dormant sprays for control of pandemis leafroller reduced damage by this pest in problem orchards. During 1989 and 1990, control failures involving chlorpyriphos have been reported with increasing frequency.

Obliquebanded leafroller has most often been reported in high numbers in newly planted apple orchards or orchards just beginning to bear fruit. It is usually uncommon in apple orchards which receive regular insecticide cover sprays for codling moth. However, obliquebanded leafroller has been increasing as a problem of bearing orchards in the Babcock Ridge area near Quincy, Washington, in the Mattawa, Washington, area, and has been the primary leafroller problem reported from the Milton-Freewater area of Oregon. The fruit tree leafroller and European leafroller have not been reported as pests of significance in Washington State orchards, although the former is a problem in orchards of British Columbia.

IDENTIFYING CHARACTERS

Adults

The adult pandemis leafroller is a buff- or tan-colored moth one-half to three-quarters of an inch in length. The main distinguishing characteristic of this moth is the banding pattern present on the wing. When the moth is at rest, a darker tan or brown band runs obliquely across the middle of each wing, appearing to join at the center. On the leading and trailing edge of this darker band is a narrow, light tan to yellow line separating the darker area from the lighter colored portion of the remainder of the wing. Color intensities vary considerably from individual to individual.

The obliquebanded leafroller moth is similar in appearance to the apple pandemis but is generally larger, three-quarters to one inch in length, and darker. The banding pattern on the wing is similar, but the leading and trailing edge of the band extending across the wing is darker than the remainder of the wing, not lighter as in the apple pandemis.

The fruit tree leafroller adult has a less well-defined pattern of banding on the wings and is easily differentiated from the two former and the next species. There is considerable variation in the color intensity of wings. However, the overall color is a golden brown with the dominant character being the presence of two cream-colored spots on the bottom (leading edge) of each wing. The fruit tree leafroller moth has a much more mottled appearance when compared with the other species.

The European leafroller adult is about the same size as the pandemis leafroller but a darker chocolate brown in color. The banding pattern on the wing is similar to that of the obliquebanded leafroller, but the edges of the darker bands are dark brown.

Eggs

Eggs of the pandemis leafroller and obliquebanded leafroller are laid in masses of 50 to 200 on the upper surface of leaves. When initially laid, the egg mass is a light green color, but as the eggs mature, the mass turns light brown, and just prior to hatch, the dark head of the small larva can be seen in each egg. After hatch, egg masses often remain on leaves and are much more visible, appearing white on the dark green background of the leaf.

Eggs of the fruit tree leafroller and European leafroller are laid on smooth bark of young wood. The masses contain about 60 eggs and are covered with a gelatinous substance, so the outer surface appears smooth and individual eggs cannot be seen. After the egg mass has hatched, it has a perforated appearance due to the presence of larval exit holes in the surface.

Larvae

It is difficult to determine the leafroller species present in an orchard when only young larvae are present. Characteristics of older larvae are more stable and distinctive, and species can be more easily determined. Leafroller larvae will most often be found either inside a leaf which has had one side folded and attached to the other, between two leaves attached together, or beneath a leaf attached to an apple with webbing. When disturbed, larvae wiggle rapidly backwards, often dropping from the shelter on a slender thread.

The young larva of all four leafroller species have a green body, but the head capsule and thoracic shield (the segment just behind the head capsule) may range in color from light brown to black. Older leafroller larvae, the fourth and fifth instars, can be distinguished by comparing colors of the head capsule and thoracic shield. All have a green body. The head and thoracic shield of the pandemis leafroller is usually the same color as the body, light green, but may range in color from green to light brown. The obliquebanded leafroller larva has a brown to black head, while the thoracic shield can vary from brown to a dull green color. The sides of the thoracic shield are always dark brown to black. The head and thoracic shield of the fruit tree leafroller are dark brown to black. The leading edge of the thoracic shield often is white or cream-colored. This description does not agree with that given in Chapman and Link's book on the tortricid fauna of apple in New York but has been consistent with observations by this author in Washington State. The larva of the European leafroller has a moderate to dark brown head, while the thoracic shield varies from greenish brown to dark brown.

Pupae

The pupae of all four leafroller species are typical of a moth chrysalis. When first formed, the pupa is a light green or greenish-brown color, turning completely tan after a short period of time, and later a darker brown. The pupa is found in a protected place, usually a leaf which has had the sides folded and webbed together. The pupae of all leafrollers encountered in an orchard appear very similar, thus are not helpful in determining species. However, holding the pupa in a container for a few days until the moth appears makes species identification easier.

LIFE HISTORY OF LEAFROLLERS

Two-Generation Leafrollers

The life history of pandemis leafroller and obliquebanded leafroller are very similar and are summarized in Figure 1. Both leafrollers overwinter as young larvae (second or third larval instar) within a hibernaculum. The hibernaculum is composed of silken threads and is located in protected areas of the scaffold limbs (e.g., pruning scars

FIGURE 1

Generalized life history of the pandemis, *Pandemis pyrusana*, and obliquebanded, *Choristoneura rosaceana*, leafrollers.

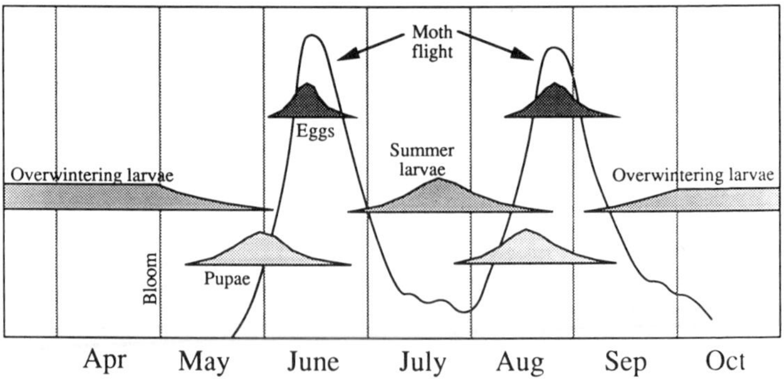

PANDEMIS AND OBLIQUEBANDED LEAFROLLER LIFE HISTORY

or small crevices in the bark). Young larvae of the overwintering generation become active in the spring as fruit buds begin to open. Most larvae have left their hibernacula by the half-inch green stage of apple bud development. These young larvae bore into opening fruit buds and later feed on expanding leaves and flower clusters. Larvae are fully grown by mid to late May. Pupae are present from mid May through early June. First adults of the overwintering generation can be detected from late May to early June, depending on spring temperatures. Peak adult activity occurs from early to mid June.

Egg hatch of the summer generation varies from year to year but maintains the same relationship relative to adult emergence. Peak adult activity of the summer generation occurs in mid to late August. Eggs of the overwintering generation are deposited in mid August.

Hatch begins in late August, continuing through September in some years. Young larvae feed for a short time on foliage and fruit, before moving to scaffold limbs where they construct hibernacula in October. Leafroller larvae found in orchards in July and August are most likely either pandemis leafroller or obliquebanded leafroller.

Single-Generation Leafrollers

The life histories of the fruit tree leafroller and European leafroller are similar and are summarized in Figure 2. Both leafrollers overwinter in the egg stage. The egg masses occur on smooth bark of one- to three-year-old wood. Eggs hatch begins at about the half-inch green stage of flower bud development and continues to bloom. Larvae feed on leaves and flowers parts and, when present, on

FIGURE 2

Generalized life history of the fruit tree, *Archips argyrospilus*, and European, *Archips rosanus*, leafrollers.

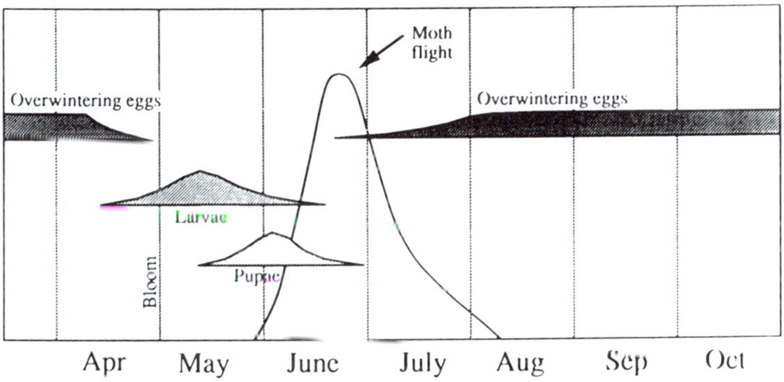

FRUIT TREE AND EUROPEAN LEAFROLLER LIFE HISTORY

young fruit. Larvae are fully grown by mid to late May. Pupae are present from late May through early June. First adults can be detected from early June. Peak adult activity occurs in mid June, and the flight is usually over by early July. Egg masses are deposited on bark where they remain until the following spring. There is only one generation each year.

DAMAGE CAUSED BY LEAFROLLERS

Larvae feed primarily on foliage, which has no economic impact on mature trees unless populations are extremely high. The primary economic impact occurs when larvae feed on fruit. Fruit feeding occurs when larvae inhabit foliage close to or touching fruit or when fruit occurs in clusters. Such situations increase the chance of a larva attaching a leaf to an apple or using fruit clusters as a sheltered feeding site. Larvae, however, do not appear to seek out fruit as a favored feeding substrate.

Larval feeding damage to fruit by all leafroller species in May to early June appears very similar. Larvae feed on fruit in clusters consuming portions of young fruit. If these fruits do not abort, the feeding damage leaves them deeply scarred and often severely deformed. This is the only time fruit tree leafroller and European leafroller can damage fruit. However, pandemis leafroller and obliquebanded leafroller can cause fruit damage two more times during the year.

The second period when fruit damage occurs is during late June and early July, when young summer generation larvae attach a leaf to the apple surface and feed on the apple. Damage at this time is characterized by shallow feeding, usually of less than one-sixteenth of an inch deep. Feeding areas usually are not contiguous but occur as several separate feeding sites within a small area. As larvae grow, they tend to move from the interior fruiting canopy and feed on foliage of growing shoots. However, more extensive feeding damage can occur during July if larger larvae continue feeding on fruit. This type of fruit damage may be more common on varieties like Granny Smith and has usually been observed where obliquebanded leafroller is present. Feeding sites made at this time of year have scabbed over by harvest and are often associated with a lighter colored area on the fruit where a leaf had been attached.

The third period during which feeding damage can occur is just prior to and during harvest. Larvae of the overwintering generation emerging from egg masses in late summer may feed on fruit, usually beneath an attached leaf. This damage is often difficult to detect. The young larva chews a small hole through the skin of the apple and then feeds on a small quantity of apple flesh directly beneath the entry site. Two or more of the "pinhole" feeding sites may occur within a small area. While this damage is difficult to detect at harvest, after being in storage, decay or drying of apple flesh around feeding sites makes them much more visible. This type of damage is often misidentified in the packing line as codling moth damage.

The distribution of leafroller larvae and their associated damage depends on tree size and type. In large trees, two to four times more larvae may occur in the upper half compared with the lower half. Spur varieties tend to have more fruit damage than non-spur varieties at the same larval densities. The closer association of fruit and foliage in spur varieties increases the likelihood of leaves contacting fruit, thus providing suitable larval feeding shelters for the pandemis leafroller. Much more research is required on the relationship between leafroller larval densities and fruit damage. Data relating densities of overwintering pandemis leafroller larvae to fruit damage in early June indicate that for every 10 percent fruit bud infestation in the prebloom period there will be about one percent fruit damage on spur Red Delicious. This relationship will be different for other varieties, especially non-spur types.

Pandemis leafroller and obliquebanded leafroller have been reported infesting sweet cherry orchards. Young larvae have been detected inside fruit and have been mistaken for cherry fruit fly larvae until examined more closely. Reduced spray programs after cherry harvest allow development of pandemis leafroller and obliquebanded leafroller, making it possible for large populations to build up and affect nearby apple orchards. Some severe damage to cherry from obliquebanded leafroller has been reported in The Dalles, Oregon.

LEAFROLLER DEVELOPMENT

Rearing of pandemis leafroller in the laboratory indicates that the lower developmental threshold is 41°F. The upper threshold has been less well characterized but falls between 80 and 85°F. These developmental thresholds are similar to those reported for the obliquebanded leafroller, 43 and 85°F. Eggs of pandemis leafroller require seven to 10 days to hatch at temperatures between 80 and 90°F. Pandemis leafroller larvae require an average of 70 days to develop at 54°F but only about 40 days at 80°F. Pandemis leafroller larvae destined to become males complete development faster than those destined to be females, when reared at the same temperatures. However, male and female pupae develop at the same rate in the same temperature. Pandemis leafroller pupae require about 11 days before emerging as moths at 62°F, but only seven days at 78°F.

A model predicting the development of pandemis leafroller is in the final stages of research and is being field tested at several locations in 1991. The essential components of the model are given here but should be only used as rough guidelines until the final version of the model is available. Table 1 shows two sets of degree-day values. The column headed CM-D° is based on the degree days as calculated in the codling moth model where the thresholds for development are 50 and 88°F, and a horizontal cut-off for the upper threshold is used to determine degree-day values using the sine-wave method. The column headed PLR-D° is based on degree-days using the thresholds of 41 and 85°F and using a vertical cut-off for the upper threshold. The comparison between the codling moth and pandemis leafroller degree days may be helpful for growers who are familiar with the codling moth model. In most years, it may be possible to use codling moth degree-day values to obtain reasonable predictions of pandemis leafroller life history events and to time control or sampling activities.

The start of the first flight has varied from May 20 (1987) to June 9 (1984). It is difficult to determine when the first flight activity is complete using pheromone traps. The peak of the flight usually occurs within 10 to 14 days (June 10 to 25) following the beginning of emergence. However, moth capture continues throughout most of

TABLE 1

Degree-day values associated with key life history events of the pandemis leafroller.

Life history event	CM-D° (from CM BIOFIX)	PLR-D°[1] (from PLR BIOFIX)	PLR-D°[1] (accum. from March 1)
First moth (BIOFIX)	275	0	930
50% emergence	450	230	1160
95% emergence	650	500	1430
First egg hatch	610	420	1280
50% egg hatch	655	485	1345
First moth (2nd flight)	1500	1320	2250
50% emergence	1870	1820	2750
95% emergence	2215	2370	3300
First egg hatch	2000	1895	2870
50% egg hatch	2070	2050	3025

[1]Degree days calculated using a lower threshold of 41°F and an upper threshold of 85°F. The sine-wave method was used to calculate degree-days using a vertical cut-off for the upper threshold.

July. The start of the second flight period is difficult to determine because moth activity usually does not drop to zero. The only indication of the second flight is a consistent increase in capture of moths over a two-week period. Degree-day values should be helpful in determining if variations in moth capture represent a normal erratic pattern of activity between the first and second flight or if the increase is truly the second flight.

Egg hatch occurs over a relatively short period. In 1983, hatch of the summer generation started on June 12 and was complete by June 26. In 1984, hatch did not begin until July 3. In 1984, hatch of the overwintering generation began on September 7 and continued into early October.

CHEMICAL CONTROL OF LEAFROLLERS

Insecticides have been the primary tactic used to control leafrollers in commercial orchards. Most of the research conducted in Washington State has been directed at the control of pandemis leafroller, but it is thought that recommendations for pandemis leafroller will apply to obliquebanded leafroller control as well.

There is considerable variation in the efficacy of products against pandemis leafroller. Table 2 shows the range of efficacy of different insecticides that were tested for leafroller control between 1979 and 1985. Some insecticides were effective against only the adults, while others were effective against both adults and larvae. Several insecticides provided only poor control and have not been recommended as chemical controls for pandemis leafroller. Note that some of the products that have shown good control are not registered for use on apple or are highly toxic to predatory mites, so are not recommended.

In 1990, chemical control tests were conducted at two locations in Washington State. The timing was at the half-inch green tip stage of apple bud development, the optimum timing for control of the overwintering larvae. Only 75 percent control of pandemis leafroller larvae was achieved with Lorsban®, a product that in the past has provided excellent control. It is possible that at this time even the best chemical controls will, in some locations, offer only fair control of pandemis leafroller.

Resistance Development in Leafrollers

Worldwide, several leafroller species known as pests of tree fruit crops have developed resistance to organophosphate insecticides. Obliquebanded leafroller has been reported as having developed resistance to some organophosphate insecticides in the eastern United States. The tufted apple bud moth, *Platynota ideausalis*, is

TABLE 1

Summary of pandemis leafroller control experiments with various insecticides conducted between 1979 and 1985.

Chemical	Formulations tested	Rates lb ai/A	Percent mortality	Comments
Guthion	50WP	0.67-1.0	76	poor control at low rates and when used against larvae, fair at high rate against adults.
Parathion	25WP 4F	1.0-2.0	72	fair against larvae prebloom
Penncap-M	2F	1.5-2.0	95	good against adults in June or larvae in July
Lorsban	4E 50WP	1.0-2.0	95	good against young larvae or adults
Methomyl	90SP 1.8L	0.9-1.8	87	poor at low rate, good against larvae at high rate; kills predatory mites.
Dimethoate	2.67EC 2.4E	1.2-1.3	29	poor control
Thiodan	50WP	1.0-2.0	39	poor control
Vydate	2E	2.0	50	poor control
Supracide	2E	1.0	10	poor control
Pydrin	2.4E	0.3	90	good control, kills predatory mites
Danitol	2.4E	0.3-0.4	51	poor control, kills predatory mites
Orthene	75SP	10.-1.5	86	poor at low rate good at high rate

resistant to several insecticides in at least seven eastern states. In New Zealand, the light brown apple moth, *Epiphyas postvittana*, is resistant to organophosphate insecticides.

Failure of several organophosphate insecticides to provide control of pandemis leafroller larvae in the field suggest that this insect is developing or, in some locations, has developed resistance to these products. In 1987, 1988, and 1990, populations of pandemis leafroller from several sites throughout Washington State were tested to determine their susceptibility to azinphosmethyl (Guthion®) and esfenvalerate (Asana®). Pheromone traps were used to capture males moths. Then, a small dose of insecticide was applied to each moth. Moth mortality was determined after 24 hours at 60°F.

Using this method, it was possible to calculate the dose that would be expected to kill 50 percent of the moths, or the LD_{50}. There were no significant differences in the LD_{50} values from populations originating in Yakima, Wenatchee, or the Tri-Cities area. The average LD_{50} for all populations was 0.150 mg/ml for azinphosmethyl and 0.0058 mg/ml for esfenvalerate. Either the pandemis leafroller populations tested are equally susceptible or resistant to these products.

A population of pandemis leafroller from an area that has received little or no insecticide use has not been tested, thus comparison of pandemis leafroller populations in commercial orchards to what might be a truly susceptible population has not been made. There is indirect evidence that pandemis leafroller, and possibly obliquebanded leafroller, is becoming more tolerant of organophosphate insecticides in several areas of Washington State.

In the early to mid 1980's, chlorpyriphos (Lorsban®) provided excellent control (95 to 100%) of overwintering larvae from applications made at the half-inch green stage of flower bud development. Growers who had had problems with pandemis leafroller control using parathion and other organophosphate products at this time received seasonal control from just a single application of chlorpyriphos. During 1989 and 1990, some of these same growers reported control failures using chlorpyriphos in the delayed-dormant period and had to apply additional control treatments during the summer period. In field tests during 1990 that compared several chemicals for control of pandemis leafroller, chlorpyriphos provided only about 75 percent control.

While application methodology (sprayer speed, gallons per acre, spray coverage) can greatly influence leafroller control results, the evidence supports the contention that pandemis leafroller is becoming much tougher to control. In some area of Washington State, obliquebanded leafroller is an increasing problem in apple orchards. This seems to be especially true in the Columbia Basin (Quincy and Mattawa) and Okanogan fruit growing areas.

BIOLOGICAL CONTROL OF LEAFROLLERS

Several natural enemies have been reported attacking leafrollers. Parasitism levels tend to be low in commercial orchards, less than 20 percent, probably the result of insecticide use. In Washington State, the most common parasite found in commercial orchards is in the genus *Apanteles* species (probably *atar*), although the specific name remains in doubt. This parasite is polyembryonic, meaning that several parasite larvae (2-15) may develop from the deposition of a single egg in the leafroller larva. For this reason, it has been targeted as a candidate for further study in an attempt to mass rear it in the laboratory for inundative releases in the field. The adoption of more intensive integrated pest management practices and implementation of control tactics that are less toxic to natural enemies increases the potential for biological control of leafrollers.

REFERENCES

Brunner, J. F. 1983. Pandemis leafroller, a pest of increasing concern to Washington fruit growers. Proc. Wash. State Hort. Assoc. 79: 119-125.

Brunner, J. F. 1988. Apple pandemis in Washington apple orchards. *Down to Earth* 44: 5-9.

Brunner, J. F. and E. H. Beers. 1990. Apple pandemis and obliquebanded leafroller. Cooperative Extension Washington State University EB1582.

Chapman, P. J. and S. E. Lienk. 1971. Tortricid fauna of apple in New York. Special Publication, New York Agric. Exp. Sta., Geneva, New York. pp. 122.

Madsen, H. F., S. A. Potter, and F. E. Peters. 1977. Pest Management: Control of *Archips argyrospilus* and *Archips rosanus* (Lepidoptera: Tortricidae) on apple. *Can. Entomol.* 109: 171-174.

Meagher, Jr., R. L. and L. A. Hull. 1986. Techniques to measure azinphos-methyl resistance in *Platynota ideausalis* (Lepidoptera: Tortricidae). *J. Econ Entomol.* 79: 1130-1133.

Newcomer, G. J. and F. W. Carlson. 1952. The leafroller moth *Pandemis pyrusana*. *J. Econ. Entomol.* 79: 119-125.

Reissig, W. H. 1978. Biology and control of the oblique-banded leafroller on apples. *J. Econ. Entomol.* 71: 804-809.

Rodriguez, J. C. 1989. The use of pheromone traps in monitoring insecticide susceptibility and resistance in a leaf-rolling tortricid, *Pandemis pyrusana* Kearfott. Master of Science Thesis, Washington State University. pp. 78.

Suckling, D. M., R. B. Chapman, and D. R. Penman. 1984. Insecticide resistance in the light brown apple moth (*Epiphyas postvittana* Walker) (Lepidoptera: Tortricidae): larval response to azinphosmethyl. *J. Econ. Entomol.* 77: 579-582.

Vakenti, J. M., C. J. Campbell, and H. F. Madsen. 1984. A strain of fruit tree leafroller, Archips argyrospilus (Lepidoptera: Tortricidae), tolerant to azinphosmethyl in an apple orchard of the Okanagan Valley of British Columbia. *Can. Entomol.* 116: 69-73.

18

THE MULLEIN BUG, *CAMPYLOMMA VERBASCI*

ROBERT F. SMITH
Agriculture Canada,
Research Station,
Kentville, Nova Scotia,
Canada

THE MULLEIN BUG IS A NATIVE EUROPEAN INSECT that undoubtedly came to eastern North America with plant material brought by the early settlers. Until agriculture intensified in the early 1900's, the bug's presence was not of significance. Although apples have been grown in Nova Scotia since the 1600's, not until the late 1930's was the mullein bug's pest status recognized. In the East Coast states, it was reported by 1915. The spread westward could have easily coincided with relocation of fruit tree nursery stock.

Many of the early pesticides were very harsh on most orchard insect life. Growers and extension specialists involved with tree fruit production soon recognized the need for long-term, in-depth research into the impact of *all* sprays on both pests and beneficials.

By 1915, considerable research was underway, involving federal entomologists stationed in Nova Scotia. These studies encompassed long-term effects of fungicides and insecticides on the total insect complex and led to recommendations promoting the least disruptive products available. A consequence of this research was a pest management approach termed "harmonized pest management," which eventually evolved into the widely recognized "integrated pest management" (IPM) of today.

These studies spanned several decades and transcended several teams of entomologists. Many valuable insights were gained from these works; the beneficial role of predators and parasites in orchards was particularly noteworthy. We also discovered there are twelve species of predaceous bug plus three pest species—mullein bug, apple brown bug, and the tarnished plant bug. Miridae is the family classification for this group, commonly termed the plant bugs. In Canada, there are more than 80 species in this family; some are strictly predators, some depend entirely on plant material food, while others use both insects and plants as a food source.

Damage Outweighs Benefits

Detailed field observations demonstrated that mullein bug do feed on many orchard pests, including European red mite, aphids, and psylla, but it remains unresolved how significant its impact is as a predator. Some people today attempt to justify this bug's presence as a beneficial predator in our orchard systems, but I personally doubt that benefits outweigh the damage caused. Furthermore, the bug cannot be trusted!

Occasionally, high populations fail to injure a crop, but who growing susceptible fruit can take the risk? If the mullein bug were a major predator that worked season-long as an ally, we would all sing its praises; but the multi-million dollar price tag for its injury to susceptible varieties should make its role as a key pest clear to everyone. Until we no longer grow these vulnerable varieties, there are limits to tolerance of this plant bug.

Most growers are familiar with injury caused by mullein bug. It is only the nymph that damages fruit. Red Delicious, Golden Delicious, Spartan, and some of the older types such as Northern Spy are prone to damage. Even those less susceptible varieties like McIntosh can be rendered culls by high populations of this bug. Corky scar tissue is a symptom of the fruit's reaction to digestive juices injected into the flesh when nymphs puncture and feed. In Nova Scotia, a significant percent of our orchards have required control measures for these pest mirids.

BIOLOGY OF THE BUG

Life Stages and Life Cycle

Mullein bugs overwinter as diapausing eggs deposited in fruit spur wood. Only the tip of an egg is exposed; their cryptic nature makes them virtually impossible to detect, even with a good microscope. The egg cap is smaller in diameter than a red mite egg. These overwintering bug eggs can withstand harsh weather; studies done at our Kentville, Nova Scotia research station indicate survival at temperatures down to minus 22°F.

We believe that these eggs are in true diapause, because an extended cold period is needed before hatch can occur later the following spring. In the Okanagan Valley of British Columbia, by mid December, eggs had sufficient cold treatment and would eventually hatch when incubated at warm temperatures for a few weeks.

There are several aspects of egg development and hatch that I studied during my two and a-half years of mullein bug research in British Columbia. As winter progressed, development within the egg advanced. Eggs brought in and incubated require a diminishing amount of time before hatch.

FIGURE 1

Advancement of the pre-hatch period in a sequence of field samples from Red Delicious apple wood infested with overwintering mullein bug eggs, Summerland, British Columbia, 1987.

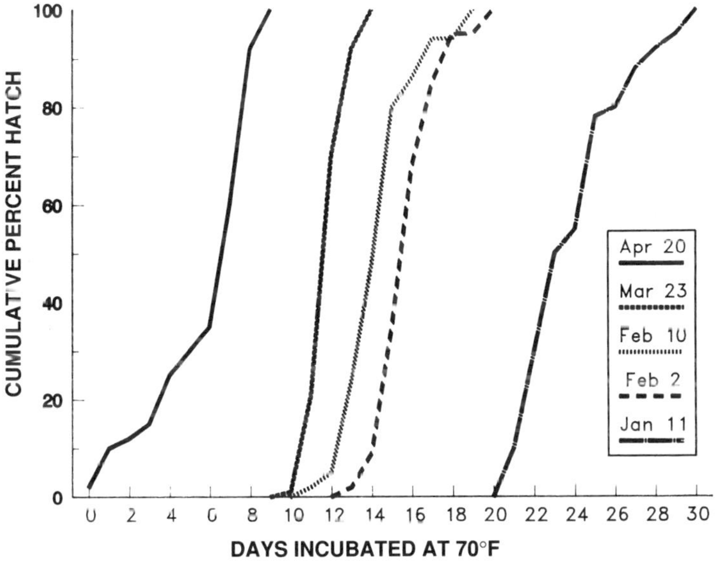

Egg Development

Often researchers attempt to predict the time of field hatch through the use of a computer-based degree-day model. There are assumptions that must be made regarding bug development if we are to predict field hatch and better time sampling schemes and sprays. One major assumption is that within a given temperature range, an incremental increase in temperature would be accompanied by corresponding advancement in development. There is a lower threshold temperature, above which accumulated daily heat units promote development and egg hatch. For the mullein bug, my early work showed hatch did not occur when overwintering eggs were held at 50°F or lower, but there was development within the egg. So, I used the 50°F as a rough hatch threshold for the model. Accumulated heat units began as of January 1 each year, and at least in the three years of my work in British Columbia, this model showed consistency in describing field hatch; the next stage is to test it in predicting hatch.

Possibly this approach would prove useful if tested for Washington State growers; I would encourage researchers there to test this program under local conditions.

The beginning of field hatch usually coincides with the bloom

FIGURE 2

Heat units based on observed field hatch of mullein bug. Thirty-one orchards were sampled in the British Columbia interior, 1986 to 1988.

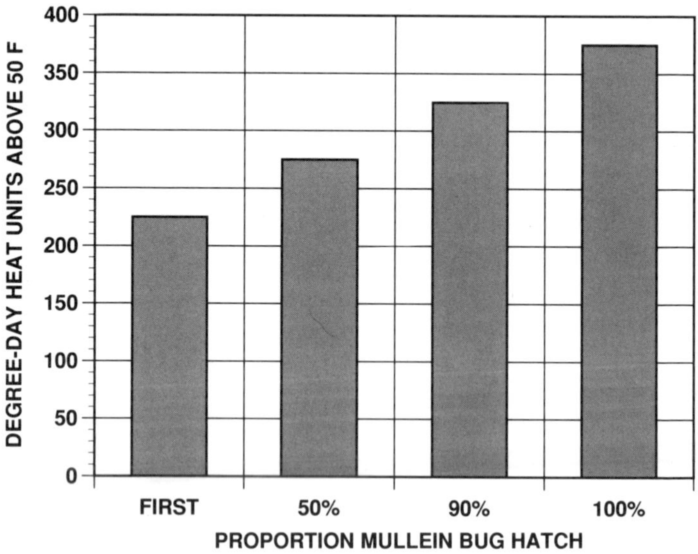

period, both in apple and pear. How long it takes for 100 percent hatch varies greatly from year to year and within any year; among growing regions there are also differences. A cool period will interrupt the process and may add several days to its duration.

TABLE 1

Field hatch of overwintering mullein bug in one Red Delicious orchard at Summerland, British Columbia, based on limb-tap samples taken at 24-hour intervals.

	Date full bloom		Days from January 1 until hatch			
Year	Calendar date	Days from January 1	First hatch	50% hatch	90% hatch	100% hatch
1986	May 15	135	140	148	150	155
1987	Apr 30	120	126	129	132	134
1988	May 6	126	119	132	137	138

This causes considerable risk when we sample orchards; low bug counts can mean one of two things. Possibly there is only a trace population present that will not merit a pesticide, or hatch has not finished and high numbers of bugs are yet to appear. To the best of my knowledge, all apple varieties serve as a host for the bug. There does not appear to be any particular preference for sensitive cultivars, nor does there exist any effect by variety on the rate of field hatch.

Fruit spur-wood is usually the most likely site for eggs to be laid, but I have found some deposited in watersprouts (suckers) if the bugs are very plentiful. Regardless of the eggs' distribution in the limbs, hatch from all compass quadrants takes place at the same rate

TABLE 2

Days from January 1, 1988 for 100 percent field hatch of overwintering mullein bug eggs from five fruit varieties, in one Summerland, British Columbia, orchard

Variety	Days from Jan. 1	No. of nymphs in sample
D'Anjou pear	129	59
McIntosh	129	85
Spartan	130	64
Red Delicious	131	57
Golden Delicious	131	142

This has useful implications when we set out to monitor bug abundance in the spring, because random sampling within the orchard and from the individual trees must give a representative picture of bug numbers. It is on this information that management decisions should be based. There is a trade-off in timing the count, because first nymphs might injure fruit before the last of the eggs hatch.

It is unlikely staff monitoring *Campylomma* would have time to check any particular orchard block repeatedly over several weeks. Realistically, a block of orchard should require sampling but once. Those of us in research must develop a practical approach to reliably sample for this pest.

Nymph Development

There are five nymphal stages, through which the mullein bug passes in quick succession. Newly-hatched *Campylomma* are tiny (1/32-inch in length), pale green, and very active; color is similar to that of young aphids and white apple leafhoppers. It is their rapid movement that helps separate the bug from these other orchard pests at calyx stage.

Insect prey is a necessity for survival of first and second instar bugs. Older nymphs can cope without prey, but seldom would their environment have no animals available as food. Both prey and plant material are pierced with the beak-like mouth and fluids sucked up as a liquid meal. On rare occasion, I have seen feeding damage caused to young fruit trees in a nursery, severe enough to kill terminal shoots. I suspect an early season aphid infestation attracted the adult mullein bugs, enticing them to linger in the trees.

Fifth instar nymphs molt into light green adults (1/8-inch in length). This color lasts but a few days, and as the bugs become sexually mature, they turn a light brown. If we flip the bugs on their back, it is evident that the females are more rounded in the abdomen and laden with eggs. Males in contrast are distinctly pointed in the abdomen.

So, at this time in the life cycle, we have adult mullein bugs in the fruit trees feeding and conditioning ready to start the next generation. Fruit trees are not the sole woody host, and in the West, wild rose and other woody plants may serve as a site for overwintering eggs. Based on my observations in British Columbia and Nova Scotia, most adults leave the woody host and seek herbaceous plants—in particular, common mullein is preferred. The rather conspicuous flower stalk seems a beacon to this plant bug. How far they move to find these hosts is unknown, but I suspect that odors from mullein help the bug home-in on its location. It is not unusual to

FIGURE 3

Mullein bug abundance on common mullein in the Okanagan Valley of British Columbia, Canada, showing three generations (arrows) in 1988.

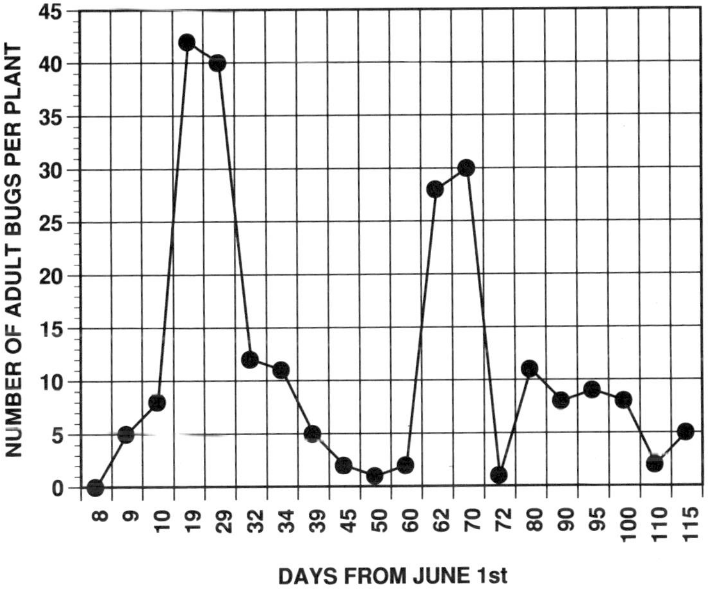

find several hundred bugs (comprised of both sexes) aggregate on a single plant. However, common mullein *(Verbascum thapsus)* is not the only herbaceous summer host; trace numbers of bugs can be found on succulent weeds and some cultivated crops. In 1988, there were three generations, one on apple and two on mullein.

For most of the summer, mullein bug feed and reproduce on herbaceous hosts. In 1986 through 1988, I established patches of

TABLE 3

Duration of mullein bug summer egg hatch from eggplant as a host.

Temperature (°F)	Number of nymphs	Days for 100% hatch*
68	95	13
72	119	10
77	145	8
81	138	7

* *Egg mortality was not determined.*

common mullein in order to study the bug's biology in greater detail. This work was supplemented by laboratory projects in which *Campylomma* were raised on eggplant. From these efforts we conclude that the female bugs live approximately 22 days (at 77°F) and are capable of producing 37 nymphs (this ranged from 2 to 184). The eggs incubated at constant temperatures took 13 days to hatch at 68°F but only seven days at 81°F.

While conducting graduate studies at Simon Fraser University, Burnaby, British Columbia, Howard Thistlewood, now a research scientist at Vineland, Ontario, proved that female mullein bugs emit a sex pheromone that strongly attracts male bugs. Less than three years after his discovery, we were fortunate enough to isolate, identify, and synthesize the lure. Prior to our breakthrough, my early work involved using caged females as a bait in pheromone traps. Even with this rather crude arrangement, I was able to determine that the males are high fliers, as demonstrated by traps that caught more at tree top heights (12 feet and above) than at lower elevations, in an orchard of standard Red Delicious.

Under field conditions, we presently cannot determine what

FIGURE 4

Captures of male mullein bugs in pheromone traps baited with caged females.

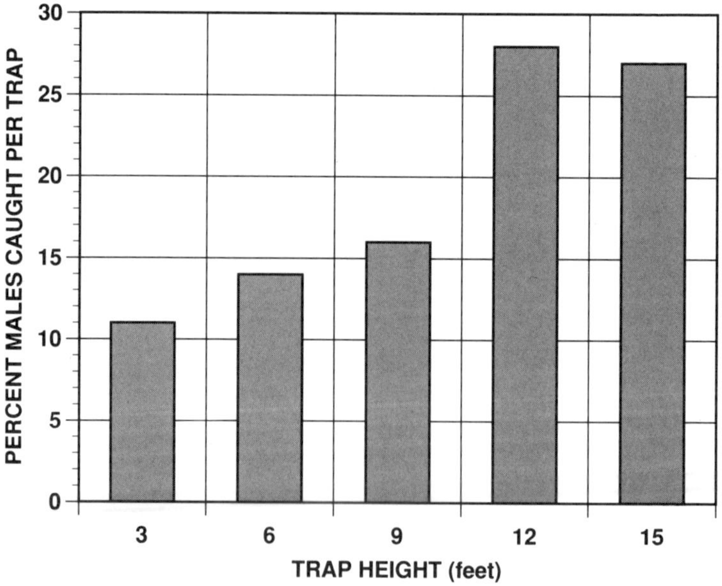

proportion of the bug population perishes, but nymph mortality through starvation or predation would be expected to influence *Campylomma* abundance. From my observations, spiders and other mirids are often the main predators; birds, too, routinely search the mullein stalks.

I believe there are normally just two full generations completed on the herbaceous plants in the British Columbia interior. Adults of the second generation migrate to woody hosts, beginning in late summer. Possibly the shorter day length triggers a response in the adults so only overwintering eggs are produced; these of course must be deposited in a woody host.

The adults that move to orchards persist well into the fall. In Nova Scotia, trace numbers can be detected through late October. In the Okanagan Valley, this flight activity could extend into November, weather permitting. Presumably, egg-laying extends over several weeks, but, regardless of the precise interval, whether an orchard will have a bug problem or not is predetermined seven to eight months in advance of the following spring egg hatch.

If fall populations could be monitored and a relationship found between those numbers of adults and nymph density the following spring, then we would have a very early warning as to which orchards should be closely sampled in the spring. I did just that in a project conducted in the British Columbia interior. Traps baited with live-caged female mullein bugs were set in commercial orchards throughout the valley, and captured males were counted weekly. The same orchards were sampled the next spring for nymphs, and statistical analysis demonstrated a consistent relationship between trap captures and nymphs. So, now we are ready to expand research directed toward a pheromone-based predictive system.

MONITORING MULLEIN BUG POPULATIONS

Action Threshold

A fundamental rule of integrated pest management is that mere presence of a pest does not necessarily mean control measures are required. Nowhere are the profits from apples so great that unnecessary sprays are affordable! Wise insecticide use also minimizes pesticide resistance and prolongs the effectiveness of registered products available to growers. Heightened concerns by our general public regarding the whole pesticide issue means that it is very much in the best interest of the tree fruit industry to project a genuine image of responsible pest management.

To resolve the presence of a consistent relationship between numbers of bugs (nymphs) and fruit injury, rather extensive research needs to be carried out in many orchard sites over at least a few years. The data would then be analyzed to determine what average number of bugs causes economic injury, that is, damage that exceeds the cost of control measures. The action threshold takes effect just prior to this, so as to avoid significant fruit injury.

TABLE 4

Action thresholds for the mullein bug based on tree fruit production guides. Total number of limb-taps are set in brackets. Each sample is approximately 20 taps per acre.

Location	Bug counts and sampling interval
Nova Scotia [20]	Cumulative count 8 sample at calyx
Ontario [250]	Cumulative 3-5 twice weekly samples at bloom plus 4 weeks
British Columbia [140]	Delicious avg. 1-10, Macs, Spartan, and pear avg. 15. Sample at bloom plus 3-day interval/3 weeks
Washington [60]	Sample at pink and continue 2 weeks through petal fall. Thresholds: Golden Delicious avg. 0.3/tray

It is generally acknowledged that action thresholds are most realistic and reliable when worked out for each geographical production area. There are lessons to be learned from what other areas are doing, but nothing compares with a developed-at-home pest management program. This, too, is a trademark of a viable IPM system—tailor-made components that explicitly deal with local factors.

Sampling Methods and Their Timing

LIMB-TAPS. What tools are available to sample and monitor mullein bugs? Since the newly hatched nymphs are the damaging stage, most efforts have focused on their detection and estimation of abundance. Jarring them onto a tapping tray is the approach, with slight variations on the theme, that has been widely recommended. Size and shape of the tapping tray have no uniform standard; this must have an effect on sample counts. In Nova Scotia, I use a square, aluminum-framed unit, 20 x 20 inches (400 in^2 in area). A good 10x hand lens is also a necessary item for bug identification.

The number of limb-tap samples per acre and action threshold are a diverse collection of values despite many of the same sensitive apple varieties being grown throughout North America.

A recent publication by Howard Thistlewood deals with a sequential sampling plan for *Campylomma* whereby as few as five limb-taps would be needed to determine if pesticides should be applied. This may be an area of promising research for Washington State, but as you might guess, it is a labor-intensive effort to get the initial background data on which growers could confidently base the reduced sampling schedule.

Determining a reliable time to sample using the limp-tap method is a major limitation to sampling and control measures. Typical field hatch will linger over a few weeks and to safeguard premature action,

TABLE 5

Sampling form using sequential decision rules and an economic threshold of two nymphs per tap, for a nominal error rate of 10%.

	Total number of nymphs	
Limb tap number	Don't treat if less than	Treat if greater than
5	2	17
6	3	20
7	5	22
8	6	25
9	7	28
10	9	30
11	10	33
12	12	35
13	13	38
14	15	40
15	17	42
16	18	45
17	20	47
18	21	50
19	23	52
20	25	54
	...stop sampling...	

If total number of nymphs is greater than 25 but less than 51 after 20 taps, the population density is not significantly different from two nymphs per tap.
Source: Thistlewood 1986, Ph.D. thesis, Simon Fraser University.

repeated samplings are recommended, such as what we see in some of the tree fruit production guides. Growers need a streamlined approach, otherwise it will be more cost-efficient, in the short-term, to apply a precautionary spray against the bug. This, of course, would be a limited gain, because insecticidal resistance, upsurgence of secondary pests, and all the other nasty side effects will come home to plague the grower. There is a degree of urgency in finding the magic formula that will enable reliable timing. A temperature-based model may be the answer; in Figure 5, I compare the predicted hatch in the Okanagan Valley with the actual observed values.

WINTER-WOOD. Some years ago, Dr. Albert MacPhee of our Kentville Research Station developed a rough-and-ready method to assess orchard blocks for damaging numbers of stinging bug. The procedure consists of collecting and incubating 20 branches each two to three feet in length, sampled one per tree. This can be done any time prior to expected field hatch, even in the cold of winter. Number of fruit clusters are counted; bugs are jarred from the limbs onto a cloth and described as counts per 100 fruit clusters. Dr. MacPhee related these numbers to fruit injury at harvest in Nova Scotia.

FIGURE 5

Actual observed field hatch at Agriculture Canada's Summerland, British Columbia, research station, spring 1988, compared to degree-day predicted dates.

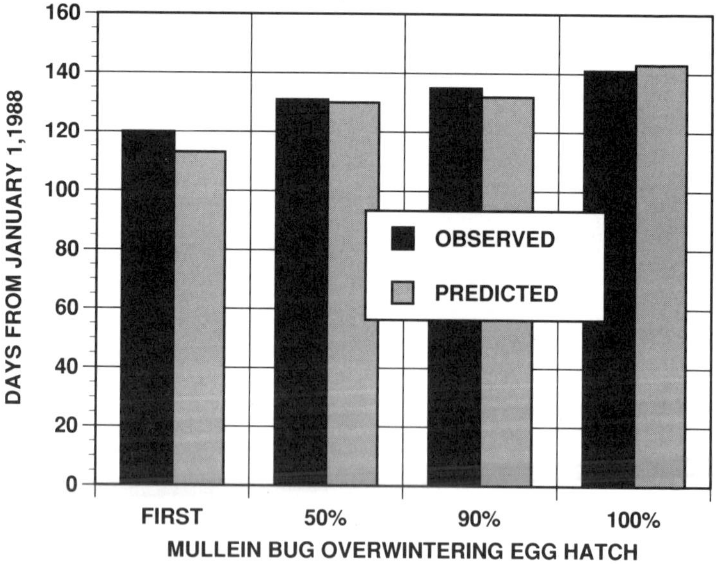

TABLE 6

Bug counts per 100 fruit clusters, Nova Scotia orchards.

10 or less	trace fruit injury	no treatment
11 to 20	approaching economic injury level. CAUTION!	possible treatment
20 or more	certain significant injury	treatment necessary

Source: MacPhee 1976. Can Ent. 108: 423-426.

PHEROMONE TRAPS AS AS PREDICTIVE TOOL. The identification of the *Campylomma* sex pheromone is the first of the plant bug family; it opens many doors of opportunity in pest management. John Borden, Harold Pierce, and associates at Simon Fraser University were instrumental in resolving the attractant. Phero Tech, Inc., Delta, British Columbia, will be the commercial developers of the synthetic lure, and presently, we are testing the product in a bubble-packet type dispenser. There are many details to be worked out before the product is ready for growers to use, and we are attempting to address these as quickly as possible.

The two-component lure is very potent, but for monitoring the orchard population of bugs, we do not want to draw males from beyond the orchard block. Compared to pheromone production of caged females, even a 0.1 percent solution of the artificial lure was a stronger attractant of males.

Trials conducted by Simon Fraser University staff in 1988 and 1989 demonstrated that fall captures using the lure relate well to subsequent spring nymph abundance. This parallels the work I had done earlier using live caged females as a bait. In the autumn of 1990, standardized experiments were started in British Columbia, Ontario, and Nova Scotia to resolve further details between male captures and bug density in the 1991 field hatch. This will put us much closer to commercialization of the product. Presently, we are using the standard wing-type trap and a bubble-packet lure release device. I envision a mullein bug monitoring system similar in many ways to that used for codling moth, but traps would be hung in late summer and captures counted at intervals throughout the fall. Field studies will be necessary for each production region, so the fine-tuned details are clearly appropriate to your orchards. Growers' bulletins like our Nova Scotia "Orchard Outlook" will play a major role in informing growers of these new developments.

TABLE 7

Number of male mullein bug captured in traps baited with synthetic lure and tested in Okanagan Valley apple orchards, September 1988.

Synthetic lure concentration	Mean number of males caught
2.0%	11.3
1.0%	8.7
0.1%	4.3
0.01%	2.0
5 live females	1.0
blank trap	0.0

CONTROL MEASURES FOR MULLEIN BUG

Until we bring a pheromone-based monitoring system on stream, the limb-tap sampling method is the best proven technique for detection and timing of pesticide application. It is far from flawless, but I would still recommend it as opposed to merely applying an insecticide at calyx stage. When significant numbers of nymphs are found, there is little spare time before treatments should be made. Depending on the product used, residual spray action may be sufficient to knock out those nymphs that hatch within several days of the spray period. We have often found that treatments applied at calyx for other pests will give some measure of bug control.

In Nova Scotia, there has been a gradual increase in the recommended rate of malathion, over the past five years. I am unsure whether this indicates a resistance problem or merely poor timing in application of this short-lived product.

Alternatives to Conventional Pesticides

At present, there are few acceptable options to those products in commercial use. Insecticidal soaps, at least from our tests, are neither very effective nor competitively priced. In 1990, we evaluated a diatomaceous earth product and found that it was reasonably useful against the stinging bugs. Unfortunately, the broad spectrum nature of this product is a little discouraging; nevertheless, in Nova Scotia, we shall be doing further trials.

The recently discovered sex pheromone of the mullein bug may hold some promise in pest management. Portions of the lure seem to repel the insect, so we will try to make use of this feature, possibly in mating disruption.

TABLE 8

Treatments applied at calyx stage of fruit development for control of stinging bugs, including mullein bug.

Region/product used	Rate product(per acre)
Nova Scotia	
1. Malathion 25 WP*	1.76 lbs.
2. Ambush 500 EC	5.71 fl. oz.
Belmark 300 EC	6.42 fl. oz.
Cymbush 12.5 WP	11.43 oz.
Decis 2.5 EC	5.71 fl. oz.
Supracide 250 EC	3.0 pints (Imperial)
3. Dimethoate 480 EC	2.4 pints (Imperial)
Cygon 480 EC	2.4 pints
System 480 EC	2.4 pints
Ontario	
Diazinon (Diazinon 50 WP)*	2.91 lbs.
British Columbia	
Diazinon (Basudin 50 WP)*	2.0 lbs.
Washington State	
Diazinon 50 WP	2 to 4 lbs.
Lorsban 50 WP	3 lbs.
Dimethoate 2.67 EC	4 pints
Dimethoate 25 WP	4 lbs.
Carzol 92 SP	1 lb.

* *Recommended treatment from tree fruit production guide.*

WHAT MORE DO WE NEED TO KNOW ABOUT MULLEIN BUG?

Despite our intensive research on *Campylomma*, there are still many unanswered questions regarding general biology. In particular, the bug's dynamics on summer hosts and cues that influence migration to fruit trees should receive further study. Mating behavior and potential of the sex pheromone to disrupt mating will be addressed by researchers in Canada. What prompts these plant bugs to attack the developing fruit? Is there a host plant volatile that triggers the response? If so, could it be masked until the apples are no longer vulnerable?

For Washington State producers, there are several vital areas of research that, in my opinion, should be addressed to improve monitoring and management of mullein bug. Some of these points may already be known or perhaps plans are underway to find the answers. a) Under your regional conditions, what numbers of mullein bugs cause an economic problem? Can an acceptable action threshold be established? b) What are the alternate hosts used by the mullein bug? What herbaceous summertime plants, near orchards, are serving as a host for the bug? Can these plants be minimized? c) Is it possible to predict spring hatch of overwintering eggs? Would a simple degree-day model consistently permit better timing of the limb-tap sampling? d) If it can be resolved when most (e.g., 90%) bugs have hatched, would a sequential sample plan similar to that developed by Thistlewood serve as an efficient way to monitor? e) Would the use of a sex pheromone monitoring system prove acceptable to growers and extension specialists as a workable part of integrated pest management?

SUMMARY AND CONCLUSIONS

The sporadic nature of the mullein bug certainly contributes to the difficulty of managing this pest. Added to this is the lack of a labor-efficient means to sample spring populations. The limb-tap approach is the best in a list of poor choices. I believe it can be made more reliable through the use of temperature-based predictive models that forecast a most likely time to sample. Furthermore, a sequential sampling plan, once tailored to your local conditions, would certainly reduce the time needed to check an orchard block. Once we have the details resolved for the use of the sex pheromone in monitoring *Campylomma,* it will be available for the necessary fine-tuning to best fit your situation.

DESIGN:
Nancy J. Born

TYPEFACE:
Weiss, set 10 on 12

PAPER:
60-lb. recycled text

PRINTING & BINDING:
Thomson-Shore, Michigan

ORDER INFORMATION

Additional copies of
New Directions in Tree Fruit Pest Management
may be ordered from:

Good Fruit Grower
1005 Tieton Drive
Yakima, Washington 98902
Phone 509/575-2315; Fax 509/453-4880

The cost per book is $10.00.
Please add $3.00 shipping and handling
for first book ($4.00 outside U.S.A.); add $1.00
shipping for each additional book.